Edited by
Ranu Jung

Biohybrid Systems

Related Titles

Wandelt, K. (ed.)

Surface and Interface Sciences

Volume 1: Basic Concepts and Methods

approx. 500 pages

Hardcover

ISBN: 978-3-527-40488-9

Schuster, H. G. (ed.)

Reviews of Nonlinear Dynamics and Complexity

Volume 3

260 pages with approx. 59 figures

2010

Hardcover

ISBN: 978-3-527-40945-7

Akay, M. (ed.)

Handbook of Neural Engineering

662 pages

2007

Hardcover

ISBN: 978-0-470-05669-1

Kumar, C. S. S. R. (ed.)

Nanodevices for the Life Sciences

489 pages with 211 figures and 10 tables

2006

Hardcover

ISBN: 978-3-527-31384-6

Waser, R. (ed.)

Nanoelectronics and Information Technology

995 pages with 1148 figures and 47 tables

2005

Hardcover

ISBN: 978-3-527-40542-8

Edited by Ranu Jung

Biohybrid Systems

Nerves, Interfaces, and Machines

WILEY-VCH

WILEY-VCH Verlag GmbH & Co. KGaA

The Editors:

Ranu Jung, PhD
Florida International University
Department of Biomedical Engineering
10555 W Flagler St., EC 2602
Miami, FL 33174
USA

Library of Congress Card No.: applied for

British Library Cataloguing-in-Publication Data
A catalogue record for this book is available from the British Library.

Bibliographic information published by the Deutsche Nationalbibliothek
The Deutsche Nationalbibliothek lists this publication in the Deutsche Nationalbibliografie; detailed bibliographic data are available on the Internet at http://dnb.d-nb.de.

Composition Thomson Digital, Noida, India

Printing and Binding Fabulous Printers Pte Ltd

Cover Design Adam-Design, Weinheim

Printed in Singapore
Printed on acid-free paper

Print ISBN: 978-3-527-40949-5
ePDF ISBN: 978-3-527-63938-0
oBook ISBN: 978-3-527-63936-6
ePub ISBN: 978-3-527-63937-3

Contents

Preface

In the last decade of the 20th century, the "Decade of the Brain", the scientific community put forth a concerted effort towards understanding the nervous system. Although experimental neurophysiological approaches provided many advances, it became increasingly evident that mathematical and computational techniques would be required to achieve a comprehensive and quantitative understanding of neural system function. "Computational Neuroscience" emerged to complement experimental neurophysiology. Simultaneously, fueled by engineering breakthroughs, the last two decades have seen a phenomenal rise in our ability to probe the nervous system and to influence neural system activity across scales of complexity and states of disease. Devices that use focused electrical stimulation to activate neural circuits are now routinely used to restore hearing to the deaf and to alleviate the symptoms of Parkinson's disease, while emerging technologies will provide amputees with the ability to feel with their artificial limb. In the first decade of the 21st century, this new engineering paradigm that links living with non-living systems to investigate, intervene and harness neural plasticity to counter disease and disablement emerged in the form of "Neural Engineering".

This book presents a window into the convergence of Computational Neuroscience and Neural Engineering. Over the past two decades it has been my privilege to be enriched by the flourishing of both Computational Neuroscience and Neural Engineering and to have the opportunity to dialogue with neuroscientists, mathematicians, physicists, and engineers from around the world. Two summers have played an important role in my personal engagement with these fields. One was a summer at Woods Hole, attending the 'Methods in Computational Neuroscience Course'. Here, I listened to John Rinzel present phase space analyses methods, talked to Ron Calabrese about leech heart interneurons that I modeled, heard about the newly devised 'Dynamic Clamp' from Eve Marder, talked about 'Consciousness' with Christof Koch and others on the beach at night, and met a neuroscientist who became my postdoctoral mentor - Avis Cohen. It was Avis who suggested a summer at Telluride at the 'Neuromorphic Engineering' workshop. There, I listened to Rodney Douglas and Misha Mahowald, once again Christof Koch, and got introduced to the world of engineers trying to capture the biological neuron in hardware. It is not surprising then, that as a biomedical engineer fascinated by the two fields, I have sought to find a practical interface that is driven by the merger of the software and hardware models of neurons with the nervous system itself. It is at the summer courses that I met many of my fellow scientists and engineers who have over the years sought similar goals, some of who have contributed to this book.

Growth of such a transdisciplinary effort required a concerted investment by many institutions that were guided by people with foresight and boldness. Dennis Glanzman and Yuan Liu from the National Institutes of Health, USA and Kenneth Whang from the National Science Foundation, USA have played an unrelenting role in supporting programmatic growth of Computational Neuroscience and the research effort of several investigators. The Collaborative Research in Computational Neuroscience Program has supported a wide range of research efforts that underlie the development of biohybrid systems and has allowed me to seek new knowledge in spinal organization for motor control after spinal cord injury. The book and I have also benefitted from transdisciplinary dialogue on biohybrid systems and neuromorphic design at a series of workshops that we conducted with support through the Science of Learning Centers program at the National Science Foundation, USA under Soo-Siang Lim. Grace Peng from the National Institutes of Health has been a steady champion of programmatic growth in neural engineering and has been a supporter of the efforts of many, including me, in bringing technology to the people that stand to benefit from this technology. Most interestingly, Elmar Schmeisser from the Army Research Office saw promise in our work on neuromorphic control of spinal interfaces in the lamprey as the basis for a novel approach to control powered or thoses for people with lower limb dysfunction. It was a presentation of these multiple related areas of research that caught the attention of Wiley and I thank them for inviting me to develop a book to present our ideas about this emerging field of biohybrid systems. The growing interest in this topic motivated my colleagues and meto develop a book for a cross-section of scientists and engineers. We hope that this book will enhance the communication between computational neuroscientists and neural engineers and bring to attention the exciting new applications that biohybrid systems could offer clinicians who are eager to deliver new solutions to their clients. It has been my pleasure to have worked with the authors of the different chapters and their teams in the writing of the book. I thank them for their effort and for their enthusiasm, not only in penning their own chapters, but also in providing helpful critiques of others.

I must thank my brother Vikram who has over the many years shared with me many of his management skills that have allowed me to juggle multiple projects and work across academic-clinical-industrial partnerships. My parents, Sarla and Padam, are a steady source of support and guidance. My husband Jimmy and son Nikhar, who are both contributors to this book, have been my sounding boards, have withstood my immersion in various projects, but most importantly have been a never-ending source of joy and companionship. Finally, I am forever indebted to my doctoral thesis advisor, Peter Katona who fostered inquiry across boundaries, supported my inquisitiveness and nurtured my foray into new realms.

June 30, 2011
Miami, Florida

List of Contributors

James J. Abbas
Arizona State University
School of Biological and Health
Systems Engineering
Center for Adaptive Neural Systems
P.O. Box 874404
Tempe, AZ 85287
USA

William Barnett
Georgia State University
Neuroscience Institute
100 Piedmont Ave. SE
Atlanta, GA 30302
USA

Guilherme Bontorin
Université de Bordeaux 1
Integration from Material
to System (IMS)
IPB, UMR CNRS 5218
351 Cours de la Libération
F-33405 Talence
France

Yannick Bornat
Université de Bordeaux 1
Integration from Material
to System (IMS)
IPB, UMR CNRS 5218
351 Cours de la Libération
F-33405 Talence
France

Rik Buschman
Medtronic Neuromodulation
Heerlen
The Netherlands

Linfeng Chen
University of Arkansas
Department of Electrical Engineering
Fayetteville, AR 72701
USA

Richard Cornelussen
Medtronic Neuromodulation
Heerlen
The Netherlands

Sharon Crook
Arizona State University
School of Biological and Health
Systems Engineering
Center for Adaptive Neural Systems
Tempe, AZ 85287
USA

and

Arizona State University
School of Mathematical and Statistical
Sciences and School of Life Sciences
Tempe, AZ 85287
USA

Gennady Cymbalyuk
Georgia State University
Neuroscience Institute
100 Piedmont Ave. SE
Atlanta, GA 30302
USA

Ralph Etienne-Cummings
The Johns Hopkins University
Department of Electrical and
Computer Engineering
105 Barton Hall, 3400 N. Charles Street
Baltimore, MD 21218
USA

Robert E. Harbaugh
Pennsylvania State Hershey
Medical Center
College of Medicine
Department of Neurosurgery
Hershey, PA 17033
USA

Ryan Hooper
Georgia Tech & Emory University
Wallace H. Coulter Department of
Biomedical Engineering
313 Ferst Drive
Atlanta, GA 30332
USA

Don H. Johnson
Rice University
Department of Electrical and
Computer Engineering
6100 Main Street
Houston, TX 77005
USA

Ranu Jung
Florida International University
College of Engineering and Computing
Department of Biomedical Engineering
10555 W. Flagler Street
Miami, FL 33174
USA

and

Arizona State University
School of Biological and Health
Systems Engineering
Center for Adaptive Neural Systems
Tempe, AZ 85287
USA

Lilian Kornet
Medtronic Neuromodulation
Heerlen
The Netherlands

Jochen Lang
Université de Bordeaux 1
Institut Européen de Chimie & Biologie
(IECB)
UMR CNRS 5248
2 rue Robert Escarpit
F-33607 Pessac
France

Kevin Mazurek
The Johns Hopkins University
Department of Electrical and
Computer Engineering
105 Barton Hall, 3400 N. Charles Street
Baltimore, MD 21218
USA

Simone C.M.A. Ordelman
University of Twente
MIRA Institute for Biomedical
Technology and Technical Medicine
Faculty of Electrical Engineering,
Mathematics and Computer Science
Department of Biomedical Signals
and Systems
7500 AE Enschede
The Netherlands

Astrid A. Prinz
Emory University
Department of Biology
O. Wayne Rollins Research Center
1510 Clifton Road NE
Atlanta, GA 30322
USA

Matthieu Raoux
Université de Bordeaux 1
Institut Européen de Chimie &
Biologie (IECB)
UMR CNRS 5248
2 rue Robert Escarpit
F-33607 Pessac
France

Sylvie Renaud
Université de Bordeaux 1
Integration from Material to System (IMS)
IPB, UMR CNRS 5218
351 Cours de la Libération
F-33405 Talence
France

Malathi Srivatsan
Arkansas State University
Arkansas Biosciences Institute
Department of Biological Sciences
Jonesboro, AR 72401
USA

Francesco Tenore
The Johns Hopkins University
Department of Electrical and Computer
Engineering
105 Barton Hall, 3400 N. Charles Street
Baltimore, MD 21218
USA

Vijay K. Varadan
University of Arkansas
Department of Electrical Engineering
Fayetteville, AR 72701
USA

and

Pennsylvania State Hershey
Medical Center
College of Medicine
Department of Neurosurgery
Hershey, PA 17033
USA

Peter H. Veltink
University of Twente
MIRA Institute for Biomedical
Technology and Technical Medicine
Faculty of Electrical Engineering,
Mathematics and Computer Science
Department of Biomedical Signals and
Systems
7500 AE Enschede
The Netherlands

Sharmila Venugopal
Arizona State University
School of Biological and Health Systems
Engineering
Center for Adaptive Neural Systems
Tempe, AZ 85287
USA

R. Jacob Vogelstein
Johns Hopkins University Applied
Physics Laboratory
11100 Johns Hopkins Road
Laurel, MD 20723-6099
USA

Jining Xie
University of Arkansas
Department of Electrical Engineering
Fayetteville, AR 72701
USA

Hargsoon Yoon
University of Arkansas
Department of Electrical Engineering
Fayetteville, AR 72701
USA

1 Merging Technology with Biology

Ranu Jung

1.1 Introduction

The most important trend in recent technological developments may be that technology is increasingly integrated with biological systems. Many of the critical advances that are emerging can be attributed to the interactions between the biological systems and the technology. The integration of technology with biology makes us more productive in the workplace, makes medical devices more effective, and makes our entertainment systems more engaging. Our lives change as biology and technology merge to form biohybrid systems.

This book describes some of the recent advances and some of the key challenges faced by engineers and scientists developing biohybrid systems that interface nerves, muscles, and machines. Modern computers have high computational capacity and high rates of internal information transfer between components; similarly, neurobiological systems have high computational capacity and high interconnectivity of neural structures. Some of the key developments in biohybrid systems have been in opening lines of communication between the engineered and the biological systems. Real-time communication between a nervous system and a device is now possible, but full and reliable integration is still far from reality. In order to achieve more complete integration, some of the key challenges in biohybrid system development are to improve the quality, quantity, and reliability of the information that can be transferred between the engineered and the biological systems.

As we move forward in developing biohybrid systems, we can leverage a second key trend in recent technological developments: technology is increasingly being designed to be adaptive in its capabilities. The breakthrough about to be achieved is to close the loop in a manner that utilizes the adaptive capabilities of electronic and mechatronic systems in order to promote adaptation in the nervous system.

Biohybrid Systems: Nerves, Interfaces, and Machines, First Edition. Edited by Ranu Jung.

1.2 NeuroDesign

The nervous system functions by generating *patterns of neural activity*. These patterns underlie sensation and perception as well as control of movement, cardiovascular, endocrine, immune, and other systems. Nonlinearities and dynamical states that span scales of physical form and time are key features of the patterns that emerge from the living nervous system. Biohybrid interfaces can be developed to (1) access these neural activity patterns, (2) influence the neural activity patterns, or (3) fundamentally alter the pattern formation mechanisms (i.e., promote plasticity) (Figure 1.1). This development can be accomplished through the process of "NeuroDesign." One aspect of NeuroDesign is that the man-made abiotic systems to access or influence the neural patterns can be devised to embody the design principles of the nervous system. Here, the fundamental structure and/or operation of the technological system are based on an understanding of nervous system function. A second aspect of NeuroDesign is the process of engineering the nervous system itself. The concept here is a deliberate approach to mold and modify the structure and function of the nervous system to obtain a specific objective. In the short timescale, this can be thought of as "influence" or control of neural system function, in the medium timescale as "adaptation," and in the long timescale as "plasticity or learning" of the nervous system. In closing the loop between the nonliving and the living, NeuroDesign also allows us to merge technology and science. This merger opens new opportunities for use of technological innovation for scientific investigation and a continuous modulation of biological activity to achieve desired function.

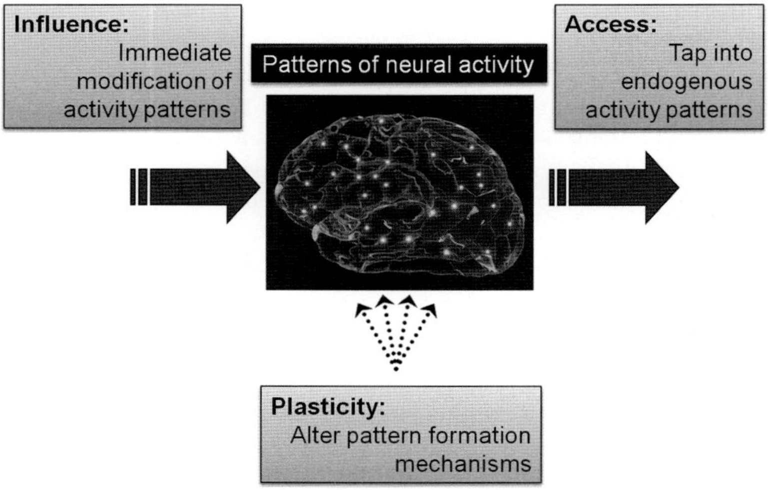

Figure 1.1 Biohybrid systems can access the patterns of neural activity, influence this pattern in real time, and induce plasticity by altering the pattern formation mechanisms. Brain image from http://www.getfreeimage.com/image/77/human-brain-and-neuron-impulses.

The primary challenge is to design biohybrid interfaces that can access and capture the biosignatures of the living system through limited spatiotemporal sampling and influence the inherently adaptive biological system through punctate intervention. For promoting plasticity, the challenge is to promote learning by influencing the core biochemical machinery in a desired manner.

1.3 The NeuroDesign Approach

Figures 1.2 and 1.3 illustrate the approach to NeuroDesign. The three features of this approach are (1) integration between the exogenous human designed system and the endogenous living system (2) biomimicry in the design of the exogenous system, and (3) the fact that an intervention that exerts its direct influence at one scale has an overall effect that spans multiple scales. The exogenous system performs both neurosensing and neuroactivation. By designing engineered systems that are biomimetic, we are able to produce systems with some of the robustness and versatility of biological systems and that potentially facilitate functional integration with the endogenous biological system. The nature and degree of biomimicry that

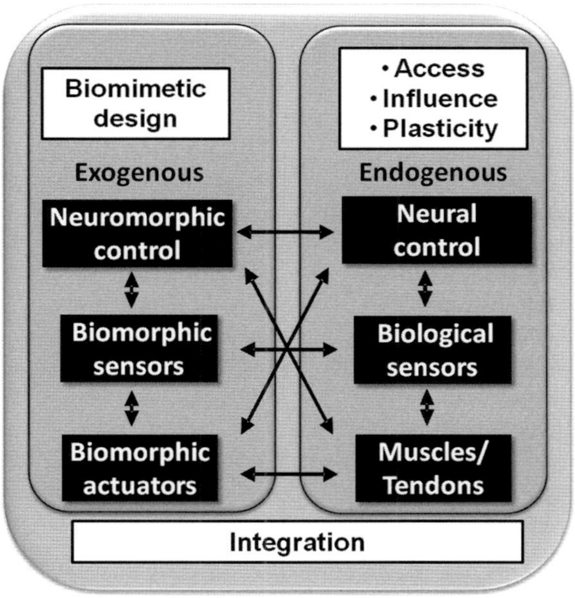

Figure 1.2 "NeuroDesign" integrates man-made systems with biological systems to access information, influence the activation of the biological system in real time, and/or promote long-term plasticity in the biological system. Bidirectional communication at multiple points of interface offers opportunities for closed-loop control of coadaptive systems. Biomimetic approaches are often used in the design of the exogenous system.

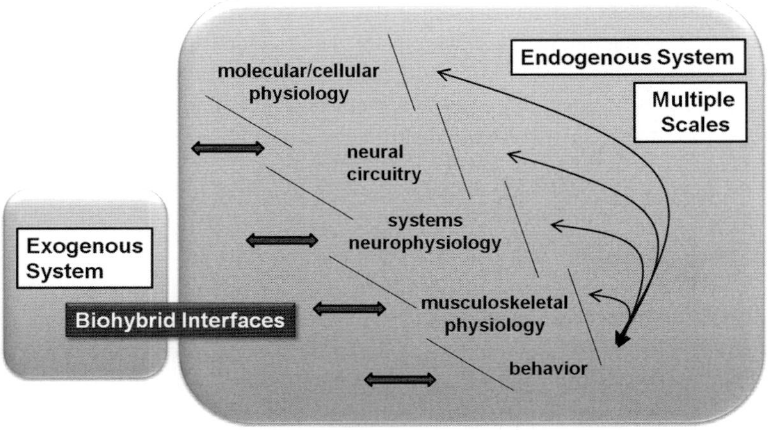

Figure 1.3 Biohybrid interfaces between exogenous man-made systems and endogenous biological systems can occur at one or more junctions along multiple scales of form and complexity. The effects of the interface at any one scale are propagated along the chain of scales.

could be used in the design of the exogenous system depend on the objective for which the biohybrid is developed. That is, when using a closed-loop system to discover ion channels at the cellular level, neuromimicry at the cellular level leads to utilization of computational models of neurons with details of ion channels. On the other hand, the development of systems for closed-loop rhythmic control of the neuromusculoskeletal system utilizes the concept of pattern generators in the nervous system to design the exogenous system.

Biohybrid systems can effect outcomes at multiple scales, at the behavioral scale (function), electrophysiological scale (synaptic learning), morphological scale (form), or molecular scale (genes/proteins/sugars). An interface that acts at one scale influences the entire chain (Figure 1.3). Thus, changes brought about at the molecular microlevel affect the pattern of activation across scales and ultimately influence behavior on a macroscale. On the other end, intervention at the macroscale for, for example, electrical stimulation of peripheral nerves after incomplete spinal cord injury to provide repetitive movement therapy, can promote motor recovery perhaps by promoting neuroplasticity at the molecular level [1–4].

Biohybrid systems can thus facilitate investigation of the intact and diseased living systems to efficiently replace damaged biological systems and to effectively interact with the residual biological components with the promise of repair.

1.4 Neuromorphic Control of a Powered Orthosis for Crutch-Free Walking

The use of NeuroDesign in the deployment of biohybrid systems can be illustrated by the following example of a powered orthotic and prosthetic system that is driven by a

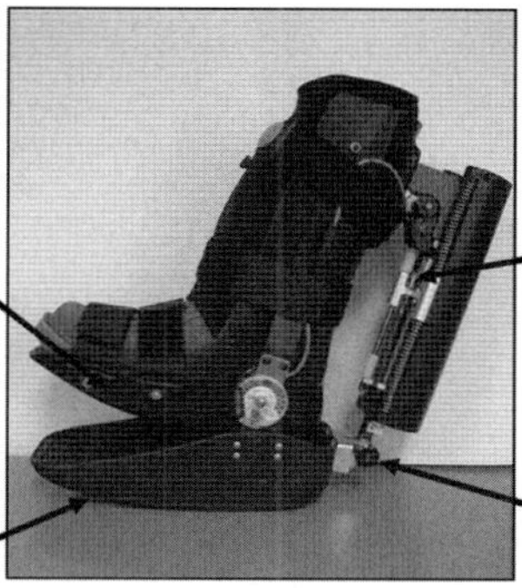

Figure 1.4 Prototype of a fixed universal ankle–foot orthosis (UAFO) attached to an AAFO. The prototype device is designed for use by combat troops. Quick release pins on the top and bottom can be used to easily separate the actuator from the AAFO.

neuromorphic controller that was designed using biomimetic NeuroDesign principles [5]. This biohybrid system (patent pending) is designed to allow "crutch-free" walking by a person with a tibial fracture of the lower limb. For this system, two objectives must be met: (1) the injured lower limb must be stabilized; and (2) the person must be able to walk under voluntary control. To achieve the former, the orthotic system illustrated in Figure 1.4 was designed. This device consists of a fixed-ankle orthosis that is used to stabilize or immobilize the injured lower limb. The fixed-ankle orthosis is encased by an actuated (powered) false-foot orthosis and the combined device forms an actuated articulated false-foot orthosis (AAFO). This AAFO is designed to permit the person to walk with a stabilized lower limb with minimal load bearing on the injured limb.

In order to achieve the second objective and provide voluntary control of the false foot, it was necessary to access information about the intent of the person to walk and then appropriately control the cyclic movement of the AAFO during walking. The inspiration for the design of this control system scheme was drawn from the control of movement in biological systems. Networks of neurons in the spinal cord of vertebrates are capable of producing rhythmic neural output that in turn controls a well-orchestrated sequence of muscle activation for cyclic control of locomotion [6]. The activity of these spinal pattern generators is usually initiated and terminated by descending voluntary control signals from the brain. The pattern generators also receive feedback from sensors in actuated muscles and tendons during the entire gait cycle. The neural organization of this biological system was mimicked in the design of the control system used for the AAFO.

An electronic circuit was designed to implement a neural network pattern generator that could be used as the controller (Figure 1.5). The biomimetic architecture of the pattern generator circuit was based on knowledge of connectivity of neurons within the spinal cord of the lamprey, a primitive vertebrate [7, 8]. Computational models of individual neurons were implemented in a circuit made from analog very large scale integrated (aVLSI) components and discrete electronic components [9, 10]. This pattern generator is capable of autonomously generating

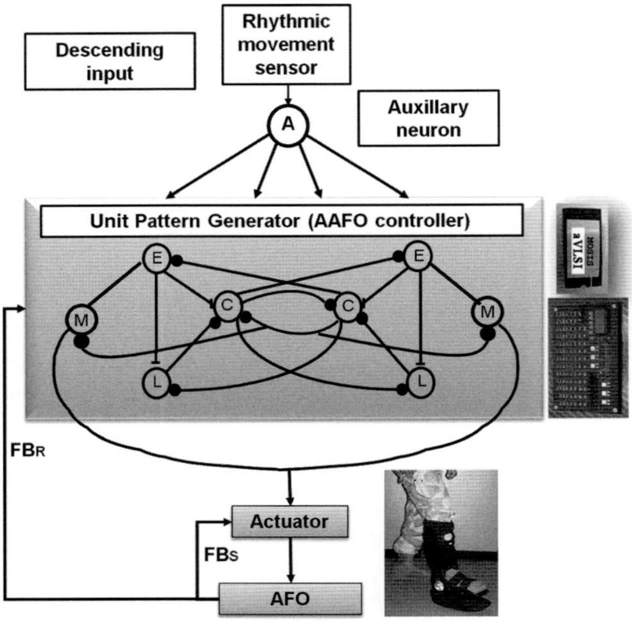

Figure 1.5 Biohybrid neuromorphic orthotic control system. The rhythmic movement sensor captures intent to move and provides periodic descending signals to entrain the unit pattern generator controller, which provides the cyclic voltage output needed to actuate the AFO. Sensors on the AFO provide local feedback (FB_S) to the actuator for control of position and ankle stiffness; sensors also provide input to the pattern generator where it may reset the rhythm in the presence of perturbation (FB_R).

cyclic voltage output that drives the AAFO. Biological pattern generators can be entrained by impinging cyclic rhythms. Their rhythm can also be reset if a perturbation of sufficient strength is applied at a particular phase of the rhythm. For example, the spinal pattern generator of the lamprey can be entrained by mechanosensory signals as well as reset by perturbations to stop and start anew [11]. Sensors mounted on the leg or AAFO provide cyclic input to the electronic pattern generator controlling the AAFO. In this manner, voluntary control of gait initiates and terminates cyclic actuation of the AAFO. Once initiated, the cadence of the AAFO matches the user's self-selected walking speed. Sensors mounted on the AAFO also provide two types of feedback signals. One set of signals feeds back position information to the actuator of the articulated ankle for local control, while another set of signals feeds information on external perturbation to the pattern generator and resets the cyclic control of the AAFO.

The importance of having an actively controlled AAFO instead of just a passively controlled ankle–foot orthosis (AFO) becomes apparent during walking (Figure 1.6). When operating in passive mode (without active control), the false foot dorsiflexes during stance phase (at approximately 40% of the gait cycle) and does not actively plantar flex at the ankle during push-off (at approximately 60% of the gait cycle). With

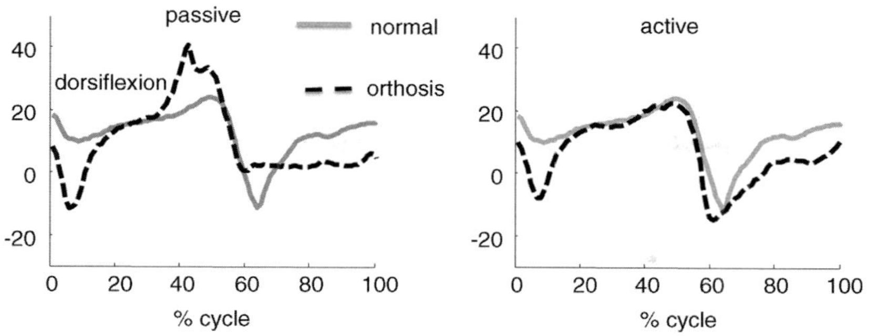

Figure 1.6 Ankle kinematics (in degrees of dorsiflexion) during a typical normalized gait cycle (heel strike to heel strike) with no orthosis (normal), passive orthosis (left), and active AAFO (right). Active control of the orthosis corrects the excessive dorsiflexion during stance phase (at 40%) and provides more plantarflexion at push-off (at 60%).

active control that is automatically timed by the entrained pattern generator, this dorsiflexion is prevented and the ankle more closely follows the normal ankle movement pattern.

Thus, this example shows how a neuromorphic design of a control system for a powered orthosis can function as a biohybrid device at the macroscale. It offers "crutch-free walking" to a person with an injured lower limb.

1.5 Frontiers of Biohybrid Systems

The greatest promise of biohybrid systems lies in promoting plasticity in the nervous system, thereby contributing to recovery and repair of lost biological function whether it ensues because of trauma, disease, or aging. This will be achieved as the closed loop becomes adaptive with adaptation occurring in both the biological and the engineered components. The greatest challenge is to design engineered systems whose adaptation enables the system to customize itself to each individual and to account for changes in the biological system as the two systems coadapt ([12–18].

As discussed and presented by multiple examples in this book, patterns of activity of the biological system could be accessed using advanced adaptive technology that responds to a biological system that is nonstationary and dynamic, and functions across multiple time- and spatial scales and multiple modalities. The design of the control system will be guided by the structural and functional constraints observed in biological systems, and allow for real-time learning, stability, and error correction that accounts for the biological systems features and takes into account the paucity of inputs to influence the biological system. The frontier lies in being able to harness the adaptive technology to promote plasticity and synergistic learning with the biological system on a long timescale under coadaptive conditions. Optimizing the technology will necessitate an approach that looks beyond the technology in isolation and looks beyond the technology as it interacts with the biological system in its current state.

Here, the design of effective technology must consider its adaptive interaction with a biological system that is continuously changing.

Endogenous compensatory learning of the biological system on short and long timescales and the physical constraints of interaction will provide challenges to this synergistic learning. It is likely that there exist windows of opportunity that may be critical periods for induction of sustained learning. Learning in the merged systems will have occurred when there are carryover effects beyond the time period when the technology is interacting with the biological systems. Future biohybrid systems may have the ability to self-wean when necessary. The biohybrid systems will thus allow us to discover the principles governing activity-dependent learning in living systems, to develop novel approaches to sense the dynamic changes in the adaptive living system and the environment, and to deliver novel adaptive technology that encourages appropriate plasticity in biological systems.

1.6
Chapter Organization

The book chapters are divided into three sections. Together, the chapters illustrate the principle approaches of NeuroDesign and present practical applications of the use of biohybrid systems for scientific interrogation and medical intervention. The first three chapters present the principles that can be used for development of biohybrid systems. Chapter 2 presents the principles of computational neuroscience. Computation complements mathematical theory and is often used to understand and reengineer the neural code represented by the rich repertoire of neural activity patterns under natural as well as experimental conditions. This chapter introduces basic physiology of neurons and presents mathematical models for excitable cells. It also presents general formalisms in neuronal modeling and briefly captures models for plasticity. The ability to embody these equivalent mathematical models for neural cells and synapses in silicon using neuromorphic electronic design principles is presented in Chapter 3. Fundamental devices and circuits that can emulate neuronal behavior at the single cell level as well as more complex circuits are presented. The chapter also discusses the advantages of using a neuromorphic approach in the design of the hardware. Chapter 4 presents principles of signal processing. It specifically examines the use of point process theory for understanding the neural code and illustrates the bounds placed by this theory in the rational design of interfaces for biohybrid systems for neurosensing and neurostimulation.

The next three chapters discuss biohybrid systems that interface at the single cell level. Chapter 5 presents the role of dynamic clamp in biomimetic and biohybrid living-hardware systems. The concepts of the dynamic clamp experimental technique are discussed and illustrated. The technique utilizes artificial synapse interfaces between single cells and computational models of those cells to investigate the fundamental biochemistry of neuronal activation. Also presented are examples of use of such biohybrid systems for specific neuronal gain control by manipulating synapses. Approaches by which the actual interface between individual neurons and sensing transducers can be enhanced by surface modification of the hardware at nanoscales that mimic biology are presented in Chapter 6. This section wraps up with Chapter 7, which introduces real-time computing for the development of the artificial neurons utilized in dynamic clamp studies. It also presents an easy-to-learn and easy-to-use technique for performing biohybrid systems analysis and presents the use of a biohybrid system to control the heartbeat in a leech though dynamic clamp.

The last section of the book consists of four chapters on biohybrid systems that interface at a macroscale and present the potential for closed-loop control of complex systems using such interfaces. Chapter 8 on biomimetic adaptive control algorithms presents the use of biomimetic features including computational models of excitable neurons, network architectures derived from biological systems, and learning algorithms inspired by synaptic learning mechanisms for the design of adaptive control algorithms. The chapter also discusses factors that should be considered in the design of closed-loop control systems in the context of coadaptation of the interfaced systems. Chapter 9 builds on Chapter 3 by presenting applications that utilize neuromorphic hardware for audition and vision and a system to control the neuromuscular skeletal system after spinal cord injury. In Chapter 10, a new approach to control cardiac function by interfacing with the nervous system is presented. It discusses the precautionary measures that will be necessary in the design of a closed-loop system. Finally, a biohybrid system with an adaptive smart sensor to measure neural activity of pancreatic cells cultured on multielectrode arrays is presented in Chapter 11. The chapter also presents the initial building blocks for a closed-loop implantable system for measuring blood-borne glucose for the management of diabetes.

Acknowledgment

Neuromorphic controlled powered orthosis case study supported by STTR Contract No. W911NF-05-C-0122 from Army Research Office to Advensys, LLC (PI: R. Jung; Contributions t case study by S.V. Jung, V. Darbhe, B. Srimattirumalaparle, D. Meller and S. Phillips at Advensys LLC and J. J. Abbas, J. Jolly and T. Sugar at Arizona State University.).

References

1 Thuret, S., Moon, L.D.F., and Gage, F.H. (2006) Therapeutic interventions after spinal cord injury. *Nat. Rev. Neurosci.*, **7**, 628–643.

2 Lynskey, J., Belanger, A., and Jung, R. (2008) Activity dependent plasticity in spinal cord Injury. *J. Rehabil. Res. Dev.*, **45** (2), 229–240.

3 Al-Majed, A.A., Neumann, C.M., Brushart, T.M., and Gordon, T. (2000) Brief electrical stimulation promotes the speed and accuracy of motor axonal regeneration. *J. Neurosci.*, **20**, 2602–2608.

4 Jung, R., Belanger, A., Kanchiku, T., Fairchild, M., and Abbas, J.J. (2009) Neuromuscular stimulation therapy after incomplete spinal cord injury promotes recovery of interlimb coordination during locomotion. *J. Neural Eng.*, **6**, 055010.

5 Jung, R., Jung, S.V., and Srimattirumalaparle, B. (2010) Neuromorphic controlled powered orthotic and prosthetic system. USPTO 201002806289 A1.

6 Gillner, S. (2006) Biological pattern generation: the cellular and computational logic of networks in motion. *Neuron*, **52**, 751–766.

7 Grillner, S., Wallen, P., Brodin, L., and Lansner, A. (1991) Neuronal network generating locomotor behavior in lamprey: circuitry, transmitters, membrane properties, and simulation. *Annu. Rev. Neurosci.*, **14**, 169–199.

8 Buchanan, J.T. (1992) Neural network simulations of coupled locomotor oscillators in the lamprey spinal cord. *Biol. Cybern.*, **66**, 367–374.

9 Jung, R., Kiemel, T., and Cohen, A.H. (1996) Dynamic behavior of a neural network model of locomotor control in the lamprey. *J. Neurophysiol.*, **75**, 1074–1086.

10 Jung, R., Brauer, E.J., and Abbas, J.J. (2001) Real-time interaction between a neuromorphic electronic circuit and the spinal cord. *IEEE Trans. Neural Syst. Rehabil. Eng.*, **9** (3), 319–326.

11 McClellan, A.D. and Jang, W. (1993) Mechanosensory inputs to the central pattern generators for locomotion in the lamprey spinal cord: resetting, entrainment, and computer modeling. *J. Neurophysiol.*, **70** (6), 2442–2454.

12 Jung, R. (2007) Adaptive learning technology. Final Workshop Report on “Future Challenges for the Science and Engineering of Learning, National Science Foundation, USA, July 23–25, 2007” (http://www.nsf.gov/sbe/SLCWorkshopReportjan08.pdf).

13 Viadaurre, C., Sannelli, C., Muller, K.-R., and Blankertz, B. (2011) Machine-learning-based coadaptive calibration for brain–computer interfaces. *Neural Comput.*, **23** (3), 791–816.

14 Gage, G.J., Ludwig, K.A., Otto, K.J., Ionides, E.L., and Kipke, D.R. (2005) Naive coadaptive cortical control. *J. Neural Eng.*, **2**, 52–63.

15 Wang, W., Collinger, J., Perez, M., Tyler-Kabara, E., Cohen, L., Birbaumer, N., Brose, S., Schwartz, A., Boninger, M., and Weber, D. (2010) Neural interface technology for rehabilitation: exploiting and promoting neuroplasticity. *Phys. Med. Rehabil. Clin. North Am.*, **21** (1), 157–178.

16 Riess, J. and Abbas, J.J. (2001) Adaptive control of cyclic movements as muscles fatigue using functional neuromuscular stimulation. *IEEE Trans. Neural Syst. Rehabil. Eng.*, **9**, 326–330.

17 Dobkin, B.H. (2003) Do electrically stimulated sensory inputs and movements lead to long-term plasticity and rehabilitation gains? *Curr. Opin. Neurol.*, **16**, 685–691.

18 Fairchild, M., Kim, S.J., Iarkov, I., Abbas, J. J., and Jung, R. (2010) Repetitive hindlimb movement using intermittent adaptive neuromuscular electrical stimulation in an incomplete spinal injury rodent model. *Exp. Neurol.*, **223**, 623–633.

2
Principles of Computational Neuroscience

Sharmila Venugopal, Sharon Crook, Malathi Srivatsan, and Ranu Jung

2.1
Introduction

The computational capacity of the nervous system is incredible. Computational neuroscience tries to capture the theoretical basis of this complex capacity at multiple scales using computational approaches. Computation complements mathematical theory and is often used to understand and reengineer the *neural code* represented by the rich repertoire of neural activity patterns under both natural and experimental conditions. A historical step in this direction dates back to the seminal works of Alan L. Hodgkin and Andrew F. Huxley who developed a neurophysiology-based mathematical model for the squid giant axon action potential [1]. Their formalism is extensively used even today, either in its original or in its reduced form. Another noteworthy contribution came from Wilfrid Rall, who emphasized the importance of the spatial domain and dendritic processing of synaptic input using mathematical approaches based on electrical cable theory [2–4]. Rall pioneered the use of digital computers in neuroscience employing a discretized version of cable theory [5], and this compartmental modeling has formed the basis for some of the most widely used software tools in computational neuroscience (e.g., GENESIS [6] and NEURON [7]). The term "computational neuroscience" was, however, coined only in the late 1980s by Eric L. Schwartz who organized a conference whose proceedings were later published in the form of a book, "Computational Neuroscience" [8]. Despite much controversy for a clear definition of the field, computational models of neuronal systems have nonetheless been successful in providing test beds for hypotheses and generating valuable neurobiological predictions. The realm of computational neuroscience has now extended beyond its original scope of complementing empirical findings and providing insights into the underlying neurophysiology. As discussed in later chapters in the book, computational neuroscience is now significantly expanding its role and influencing advancements in related fields of neuroengineering and neurotechnology. Several excellent textbooks and seminal articles have rigorously dealt with neurobiological principles and theories (e.g., [9–16]) as well as analytical tools and techniques used in computational neuroscience (e.g., [1, 2, 6–8, 17–22]).

Biohybrid Systems: Nerves, Interfaces, and Machines, First Edition. Edited by Ranu Jung.

This chapter outlines some of the predominant physiological and theoretical principles guiding neuronal modeling further indicating some of the key applications of computational models for neuromorphic and biohybrid system design and development.

2.2
Some Physiology of Neurons

Neurons and their supporting glial cells form the principal cell types of the nervous system. The *Neuron Doctrine,* as developed after the works of neuroanatomists Camillo Golgi and Ramon y Cajal, describes neurons as the structural and functional units of the nervous system [23]. Neurons form the basic information processing units of the nervous system and typically have a basic architecture consisting of a cell body (*soma*), many input processes (*dendrites*), and an output process (*axon*) (see Figure 2.1). The soma contains the nucleus and much of the cell's machinery. Each neuron usually has a single axon that arises from the axon hillock of the cell body and extends away from the soma. The plasma membrane of the axon is called the axolemma and is specialized to conduct electrical impulses originating at the axon hillock away from the neuronal cell body to other neurons. Terminal arborizations of axons end in boutons that form close appositions with the dendrites, soma, or axons of other neurons. These junctions are called *synapses.* Axodendritic synapses between the axonal terminals of the *presynaptic* neuron and specialized structures called *dendritic spines* on dendrites of the *postsynaptic* neuron are most common. Communication between the presynaptic neuron and the postsynaptic neuron typically occurs through two types of synapses, electrical or chemical. An *electrical synapse* consists of a *gap junction* between the pre- and postsynaptic neurons that allows rapid direct flow of ionic currents between the neurons bidirectionally. In *chemical synapses,*

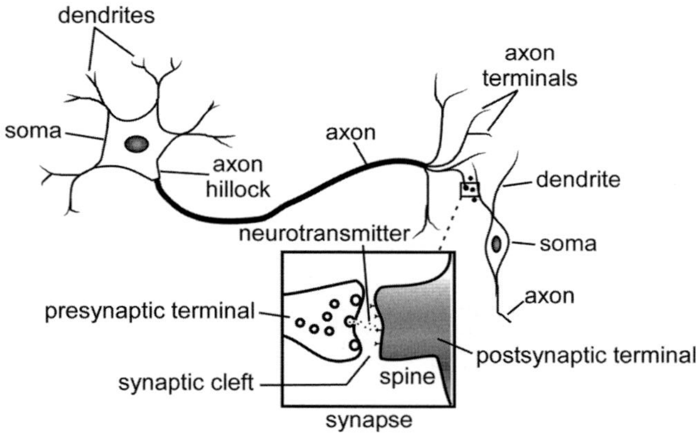

Figure 2.1 Basic architecture of a typical neuron communicating with another neuron via a chemical synapse.

the boutons at the axon terminal of the presynaptic neuron store synaptic vesicles containing *neurotransmitter* chemicals. The space between the presynaptic axon terminal and the apposing postsynaptic dendritic spine is called the synaptic cleft. Transmission of action potentials to the axon terminals of the presynaptic neuron leads to release of the neurotransmitter(s) into the synaptic cleft that consequently binds to *ionotropic* or *metabotropic* receptors (G-protein-coupled receptors) on the surface of the postsynaptic neuronal membrane. Transmitter binding to ionotropic receptors leads to opening of ion channels that result in the generation of an electrical signal in the postsynaptic neuron. Transmitter binding to G-protein-coupled receptors activates second-messenger pathways often resulting in long-term changes. Neurotransmitters can have an excitatory effect and stimulate the postsynaptic neuron to generate an electrical impulse. They can also have an inhibitory effect wherein the neurotransmitter may bind to its receptor on the postsynaptic neuron and reduce its ability to generate electrical impulses. Thus, neuronal communication and information exchange occur between neurons across synapses by means of *electrical* and *chemical* signaling.

2.2.1 Membrane Potential

All living cells including neurons maintain a *negative* electrical gradient between the inside (cytoplasm) and the outside of the cells separated by their cell *membrane*. The cell membrane encompassing the cytoplasm is a lipid (fat) barrier separating ions (charged particles) in the intracellular and extracellular fluids. Ions can be positively (cation) or negatively (anion) charged. Physiologically, the extracellular space largely consists of sodium (Na^+) and chloride (Cl^-) ions and the intracellular space is rich in potassium (K^+) and organic anions (A^-) (see Figure 2.2).

The restriction of these ions to their respective locations occurs due to electrochemical and physiological constraints. Ideally, the cell membrane is a perfect electrical insulator and hence the ions stay where they are. In particular, the larger organic anions stay in the intracellular space, conferring the neuron with a negative electrical gradient. Moreover, the differential distribution of individual ionic types across the membrane leads to their respective *chemical concentration gradients*. But the cell membrane consists of ion-selective *channels* made of complex proteins embedded within the membrane (e.g., potassium channels) that allow ions to flow across the neuronal membrane. For example, if K^+ ions were free to *diffuse* out of the cell via open K^+ channels down its concentration gradient, the negative electrical gradient of the cell prevents the outward movement of these cations. Similarly, the inward flow of the negatively charged Cl^- ions down their concentration gradient is counterbalanced by the negative membrane potential. But for Na^+ ions, both the electrical and the concentration gradients drive the Na^+ ions into the cell that could eventually neutralize the negative electrical gradient. Fortunately, at least two mechanisms prevent the inward Na^+ ion flow. The foremost is the sodium–potassium transporter (also a membrane protein, often called Na^+/K^+ pump). The transporter utilizes

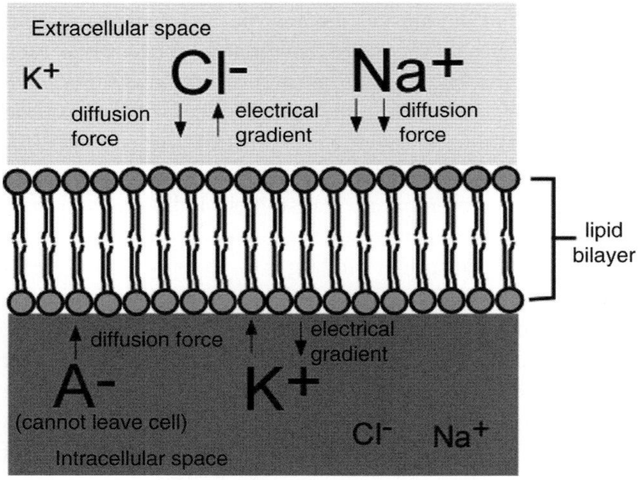

Figure 2.2 Physiological distribution of prominent ionic species across the membrane. The organic anions (A^-), potassium (K^+) ions, sodium (Na^+) ions, and chloride (Cl^-) ions are shown along with directions of the force of diffusion due to concentration gradients and electrical gradients due to differential distribution of ions across the membrane. Larger font size implies higher concentration.

cellular energy reserves (ATP) and exchanges three Na^+ ions for two K^+ ions across the cell membrane assisting in maintaining higher Na^+ concentration outside the cell. Second, in addition to the ion selectivity, most channels are sensitive to membrane potential (voltage sensitive). Hence, the voltage-sensitive Na^+ channels are typically closed at negative membrane potentials further preventing Na^+ entry into the cell.

The balance between the electrical and the chemical gradients results in equilibrium potentials for each ionic species. Thus, an *equilibrium potential* exists for the K^+ ions at which the electrochemical gradients balance each other and no net movement of K^+ ions occurs. This equilibrium potential for potassium ions can be estimated using the Nernst equation given by

$$E_K = 2.3\frac{RT}{F}\log\frac{[K^+]_{OUT}}{[K^+]_{IN}}, \tag{2.1}$$

where R is the gas constant, T is the absolute temperature and $[K^+]_{OUT}$ and $[K^+]_{IN}$ are the extracellular and intracellular potassium ion concentrations, respectively. Similarly, equilibrium potentials can be obtained for each ionic species. The combined effect of more than one ionic species on the membrane potential can be expressed by the Goldman–Katz equation as follows:

$$V_m = 58\log\frac{P_K[K^+]_{OUT} + P_{Na}[Na^+]_{OUT} + P_{Cl}[Cl^-]_{IN}}{P_K[K^+]_{IN} + P_{Na}[Na^+]_{IN} + P_K[K^+]_{OUT}}, \tag{2.2}$$

where V_m is the membrane potential and P is the relative membrane permeability for each ion type. This equation includes the contribution of any ionic gradient to the membrane potential by simply weighting its effect in accord with its membrane permeability. Note that this estimation of membrane potential is based on the assumption of constant electric field (or potential gradient) across the membrane. The resultant membrane potential is referred as the *resting membrane potential* or V_{rest} (e.g., -70 mV).

2.2.2
Membrane Equivalent Circuit

The neuronal membrane bears close correlation with electrical circuits. The lipid bilayer acts like a capacitor forming a thin insulating barrier for the ions in the intracellular and extracellular spaces. Let $V_m(t)$ describe the membrane potential difference between the intracellular and the extracellular domains at any given instant of time. The capacitance of the cell membrane (C_m) is a measure of the charge (Q) distributed across the membrane to give rise to V_m given by $Q = C_m V_m$. Current flows when the voltage across the capacitor changes; this capacitive current is given by

$$I_C = \frac{dQ}{dt} = C_m \frac{dV_m(t)}{dt}.$$

In addition, the existence of various ionic channels confers the ability to conduct charges. As noted earlier, a resting membrane displays a negative potential and resistance to the flow of charges via the ion channels. The ion-permeable channels therefore contribute to the membrane resistance, R_m. Considering a *point* representation of the neuron (ignoring the complex morphology), we can represent a resting membrane by means of a simple resistance–capacitance equivalent circuit with a battery V_{rest} as shown in Figure 2.3.

The total current flowing across the membrane (I_m) is the sum of the resistive (I_i) and capacitive (I_c) components. Applying Kirchoff's current law, the total current can be given by

$$\begin{aligned} I_m = I_c + I_i &= C_m \frac{dV_m}{dt} + \frac{(V_m - V_{rest})}{R_m} \\ &= C_m \frac{dV_m}{dt} + G_m (V_m - V_{rest}), \end{aligned}$$

where C_m is the specific membrane capacitance (F/cm^2) and R_m is the specific membrane resistance (Ω cm^2), and G_m, the inverse of R_m is the specific membrane conductance (S/cm^2). The equivalent circuit described in Figure 2.3 includes many constituent ionic currents described in Section 2.2.1. The term I_i is therefore a sum total of various individual ionic current types as follows:

$$I_i = I_{leak} + I_{Na} + I_K + \ldots$$

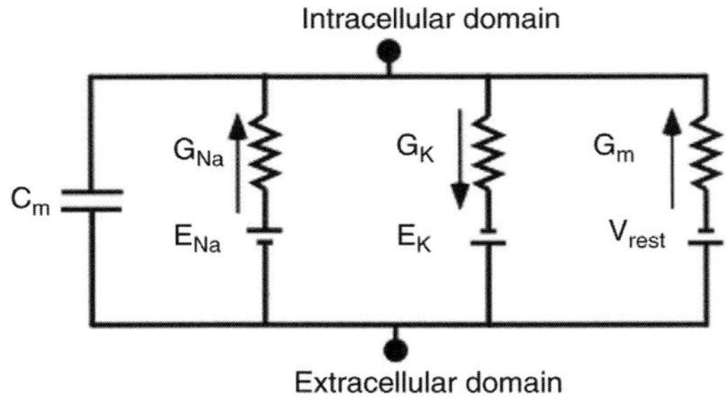

Figure 2.3 Equivalent circuit of the electrical properties of the neuronal membrane.

Here, the current I_{leak} is due to the conductance G_m and is voltage independent. The general expression for any given ionic current is given by

$$I_{ion}(t) = G_{ion}(V_m(t), t)(V_m(t) - E_{ion}),$$

where $G_{ion}(V_m(t),t)$ describes voltage-dependent ionic conductance for a given type of ionic channel; E_{ion} is the ionic reversal potential given by Nernst's equation for a particular ionic species. The *reversal potential* is the membrane potential at which a given ionic current reverses polarity. The total membrane current, I_m in the current balance equation is set to zero due to conservation of charge. Hence, the membrane current balance equation can be written as follows:

$$C_m \frac{dV_m}{dt} = \sum -G_{ion}(V_m(t), t)(V_m(t) - E_{ion}) - G_m(V_m - V_{rest}) \tag{2.3}$$

2.2.3
Action Potential: Generation and Propagation

It is possible to experimentally gain access to the cell membrane by means of sharp microelectrodes ($\leq 2\,\mu$m diameter) and introduce artificial currents to alter the membrane's potential (current clamp). Alternatively, the membrane potential can be held constant at different voltages to record the voltage-sensitive gating properties of various ionic membrane currents (voltage clamp). These two approaches have been integral to our understanding of the membrane physiology and to the development of physiologically realistic computational models of single neurons and synaptic currents.

Generation of an action potential (see Figure 2.4) involves rapid movement of ions across the membrane producing a transient change in the membrane potential leading to all or none electrical events (also called nerve impulse, electric impulse, or spike). Following from Section 2.2.1, if we inject a direct current of polarity such that it would result in de polarization of the resting membrane potential to more positive values, the depolarization-sensitive Na^+-selective ion channels open; Na^+ ions flow into the cell along their electrochemical gradient in turn making the membrane potential more positive. Increasing the magnitude of current injection will eventually

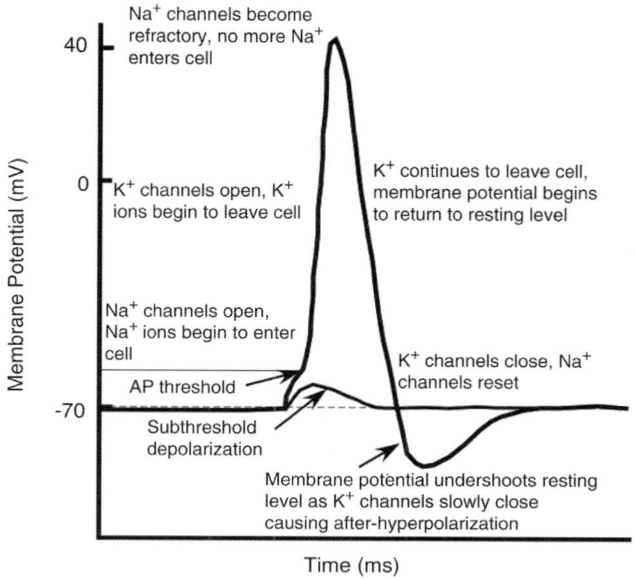

Figure 2.4 Generation of a typical action potential (AP).

allow the membrane potential to cross a certain *threshold*, resulting in an action potential (a rapid event typically lasting 1–2 ms). During the action potential, the membrane potential rapidly increases to about 40 mV (approximately the reversal potential for Na^+ ions) and Na^+ channels become *refractory* and enter an inactive state. The positive membrane potential in turn activates a set of voltage-sensitive K^+ channels called delayed rectifiers that allow K^+ to flow out of the cell along its concentration gradient. The outflow of K^+ ions often undershoots the membrane potential below resting level and the membrane potential slowly returns to V_{rest} leading to an *after-hyperpolarization* (AHP) following an AP (see Figure 2.4); the process underlies closing of the delayed rectifier K^+ channels as Na^+ channels are simultaneously deinactivated and become ready to be activated again. The refractoriness of Na^+ channels makes the generation of a second action potential impossible until the membrane is polarized to negative values close to its resting potential. Furthermore, the Na^+/K^+ pump enables extrusion of Na^+ ions that entered the cell during the action potential. Thus, the generation of action potential depends on threshold and is refractory. While the mechanisms of a classical Na^+ impulse can be thus described, action potentials of different neurons have varied amplitudes, shapes, and durations largely owing to the enormous diversity of the underlying ionic channels and their mechanisms [24].

The action potential generated at or near the soma is conducted along the axon with nearly no depreciation in its amplitude and duration. The axonal membrane resembles an electrical cable with passive electrotonic properties (known as cable properties; also see Section 2.3.3) with resistance and capacitance coupled along the length of the axon. The electrotonic currents at the site of initiation of the action potential spread along the neighboring membrane sites where, upon reaching threshold, the action potential is regenerated. The rate of spread of depolarization largely depends on the size of the axon (larger axons have higher conduction rates) (also see chapters in Refs [12, 19]).

2.3 General Formalisms in Neuronal Modeling

2.3.1 Conductance-Based Hodgkin–Huxley Model for Action Potential Generation

The seminal works of Alan Hodgkin and Andrew Huxley resulted in the first physiologically realistic mathematical model for the generation of neuronal spikes. Their model formulation was based on a series of ion replacement and voltage clamp experiments where they studied the voltage- and time dependencies of the ionic conductances underlying the action potential in a giant axon of the squid ([25–28]). Following the physiological description of action potential generation in Section 2.2.3, the Hodgkin–Huxley (HH) model consists of conductance-based formalism for the fast sodium and delayed rectifier potassium currents. In particular, they introduced fictitious gating variables that described the activation and inactivation of these currents in a voltage- and time-dependent manner. Following equation 2.3 in Section 2.2.2, the total membrane current in the HH model was given by

$$C_{\mathrm{m}}\frac{\mathrm{d}V_{\mathrm{m}}}{\mathrm{d}t} = -G_{\mathrm{K}}n^{4}(V_{\mathrm{m}}-E_{\mathrm{K}})-G_{\mathrm{Na}}m^{3}h(V_{\mathrm{m}}-E_{\mathrm{Na}})-G_{\mathrm{m}}(V_{\mathrm{m}}-V_{\mathrm{rest}})+I_{\mathrm{inj}}(t), \tag{2.4}$$

where $I_{\mathrm{inj}}(t)$ is the current injected via an intracellular microelectrode. Although the existence of ion channels was tentative at that time, the terms n, m, and h denote the activation or inactivation states of fictitious (channel) gating particles of the corresponding conductances. These terms were essentially dimensionless numbers between 0 and 1 and satisfied the linear kinetic equations as follows:

$$\frac{\mathrm{d}n}{\mathrm{d}t} = \alpha_n(1-n)-\beta_n n;$$

$$\frac{\mathrm{d}m}{\mathrm{d}t} = \alpha_m(1-m)-\beta_m m;$$

$$\frac{\mathrm{d}h}{\mathrm{d}t} = \alpha_h(1-h)-\beta_h h,$$

where α_i is the voltage-dependent rate constant (in units of s^{-1}) that specifies the number of transitions from closed to open states and β_i specifies the number of transitions from open to closed states of the gating variables. Hodgkin and Huxley derived empirical fits for the rate constants given by

$$\alpha_n(V_m) = 0.01\frac{10-V_m}{(e^{(10-V_m)/10}-1)};$$
$$\beta_n(V_m) = 0.125e^{(-V_m/80)};$$
$$\alpha_m(V_m) = 0.1\frac{25-V_m}{(e^{(25-V_m)/10}-1)};$$
$$\beta_m(V_m) = 4e^{(-V_m/18)};$$
$$\alpha_h(V_m) = 0.07e^{-V_m/20};$$
$$\beta_h(V_m) = \frac{1}{(e^{(30-V_m)/10}+1)}.$$

The above HH model reproduced experimentally similar action potentials of the squid giant axon and for the first time provided insight into the ionic mechanisms underlying all-or-none spike generation. In addition, the HH model displayed a spike threshold and refractory period (see Figure 2.5). These equations have formed the basis for conductance-based neuronal modeling.

In many instances, reduced forms of the HH model are considered to provide analytically tractable models to gain better insight into the qualitative dynamics (e.g., excitability, threshold behavior, etc.) of single neurons (e.g., see Refs [17, 18, 22, 29], and Chapter 5 in Ref. [20]). A detailed analytical treatment of reducing the four-dimensional coupled ordinary differential equations (ODEs) in the HH model to two dimensions was provided by Rinzel [30] where an earlier reduction by FitzHugh [31, 32] was also discussed and biophysically interpreted. The following two main assumptions were made to guide the reduction:

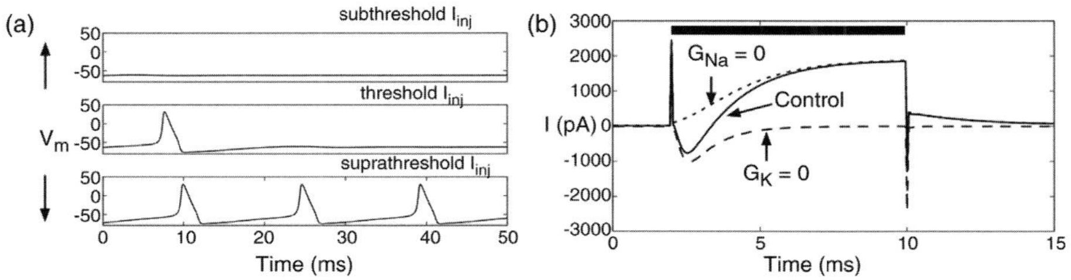

Figure 2.5 Simulations of current clamp (a) and voltage clamp (b) Hodgkin–Huxley experiments using an XPPAUT-based HH model. In (a) membrane voltage (V_m) to three levels of injected current (I_{inj}) are shown. Note all-or-none spike generation at a certain threshold I_{inj} and continuous spiking at a higher (suprathreshold) I_{inj} value. In (b) voltage clamp simulation shows (i) the composite inward (Na^+) and outward (K^+) currents under control condition, (ii) only inward Na^+ current with K^+ conductance blocked ($G_K = 0$), and (iii) only outward K^+ current with Na^+ conductance blocked ($G_{Na} = 0$). The horizontal black bar shows the duration of voltage clamp at $V_m = 0$, from a holding potential of -60 mV.

a) The activation of the Na^+ conductance is instantaneous as the observed gating characteristics of m occur within a fraction of a millisecond; this allows approximation of the temporal dynamics of m to a voltage-dependent steady state and hence eliminates the dm/dt equation.

b) A further approximation of the inactivation gating variable h for the Na^+ conductance in terms of the activation gating variable n for K^+ conductance (e.g., see Refs [17, 30] and Chapter 3) yields further reduction of dh/dt and dn/dt equations into a single ODE in terms of a new variable w that denotes the fraction of the K^+ channels in the open state.

The reduced two-dimensional ODEs are given by

$$C_m \frac{dV_m}{dt} = -G_K w^4 (V_m - E_K) - G_{Na} m_\infty (V_m)(1-w)(V_m - E_{Na})$$

$$\frac{dw}{dt} = -\frac{[w - w_0(V_m)]}{\tau_w(V_m)},$$

where $n = w$ and $h = 1 - w$, τ_w is the time constant for w. The reduced models have been invaluable in allowing the use of geometric methods and bifurcation analyses derived from dynamical systems approaches to understanding neuronal dynamics.

2.3.2
Chemical and Electrical Synaptic Inputs

Conductance-based models of neuronal excitability, based on HH formalism presented in Section 2.3.1, generally model the influence of a presynaptic neuron on a postsynaptic neuron by including a term that describes the synaptic current in the current-balance equation. The general form of the synaptic current due to a chemical synapse is

$$I_{syn}(V,t) = G_{syn}(t)(V - E_{syn}), \tag{2.5}$$

where V represents the membrane potential in the postsynaptic cell, E_{syn} is the reversal potential for the synapse, and G_{syn} is a function that describes the shape of the postsynaptic conductance. Excitatory chemical synapses are modeled using a reversal potential that is higher than the resting potential of the postsynaptic cell, and inhibitory synapses rely on a reversal potential that is lower than the resting potential of the postsynaptic cell. One common choice for $G_{syn}(t)$ is the waveform for an alpha synapse with conductance

$$G_{syn}(t) = \bar{g}_{syn} \frac{(t - t_{sp})}{\tau} e^{(1-(t-t_{sp})/\tau)} \tag{2.6}$$

for $t > t_{sp}$, where $\bar{g}_{syn}$ is the maximal synaptic conductance, t_{sp} is the time of the presynaptic spike that elicits the onset of the postsynaptic conductance, and τ is the decay time constant. Another common choice is the double-exponential synaptic conductance

$$G_{syn}(t) = \bar{g}_{syn} A e^{-(t-t_{sp})/\tau d} - e^{-(t-t_{sp})/\tau r} \tag{2.7}$$

for $t > t_{sp}$, where $\bar{g}_{syn}$ is the maximal synaptic conductance, τr and τd are the time constant for rise and decay of the postsynaptic conductance respectively, and A is a normalization factor

$$A = \frac{1}{e^{-p/\tau d} - e^{-p/\tau r}},$$

where p is the time to peak. When the presynaptic cell also is explicitly modeled by a system of differential equations, $G_{syn}(t)$ can be modeled as

$$G_{syn}(t) = \bar{g}_{syn} s(V_{post} - E_{syn}), \tag{2.8}$$

where V_{post} is now the membrane potential of the postsynaptic cell, and the term s describes how the waveform of the postsynaptic conductance depends on the presynaptic membrane potential V_{pre}.

Since electrical synapses (gap junctions) rely on channels that directly connect the presynaptic and postsynaptic neurons, a voltage change in either cell will generally cause a corresponding voltage change in the other cell. In this case, the postsynaptic current can be modeled with the term

$$I_{syn} = \bar{g}_{syn}(V_{pre} - V_{post}). \tag{2.9}$$

2.3.3
Cable Theory of Neuronal Conduction and Compartmental Modeling

Many neurons have extensive dendritic processes with complex geometry, which allows a cell to receive thousands of synaptic inputs that can interact in a highly nonlinear fashion, leading to a complex dendritic computation [33, 34]. The main purpose of the use of cable theory in computational neuroscience is to understand how these synaptic inputs interact and are integrated in the system of branching dendrites of different diameters with differing membrane properties.

When considering a long, thin neural process, which is part of a dendrite or an axon, the membrane potential varies more along the neural process than in the perpendicular dimension. By neglecting the perpendicular variation, we are able to model the variation along the long axis of a uniform dendrite with only one spatial dimension, denoted by x. With this assumption, the cable equation can be expressed as the second-order partial differential equation

$$\tau \frac{dV_m}{dt} - \lambda^2 \frac{\partial^2 V_m(x)}{\partial x^2} = -(V_m - V_{rest}), \tag{2.10}$$

where the membrane potential V_m now depends on both space x and time t. The membrane time constant $\tau = R_m C_m$, where R_m is the membrane resistance per unit area and C_m is the membrane capacitance per unit area as defined above. The space constant $\lambda = \sqrt{(dR_m)/(4R_i)}$, where R_i represents the resistivity per unit volume of the intracellular medium, and d is the diameter of the uniform neural process. For more details, a complete derivation of the cable equation is provided in Ref. [21].

We can approximate this continuous model using adjacent, connected regions of lumped, isopotential membrane by discretizing in space, which provides several computational advantages. In particular, this approximation does not require that the neural process be uniform; instead, this computational approach can be applied to dendrites of varying diameter with branching structures and nonuniform membrane properties. This is the basis for many simulation platforms developed for neuroscience, including NEURON and GENESIS (see Section 2.6).

2.3.4
Calcium and Calcium-Dependent Potassium Currents

Ionized calcium (Ca^{2+}) is a critical ionic species for membrane potential dynamics that serves as a key determinant of many different cellular/neuronal processes such as triggering neurotransmitter release during a presynaptic action potential, axonal growth, muscle contraction, and synaptic plasticity (see Section 2.4.1) to name a few. Primary mechanisms regulating the influx of Ca^{2+} into the cytoplasm include voltage-sensitive calcium channels and NMDA (*N*-methyl-D-aspartate)-sensitive channels (receptors for the endogenous neurotransmitter glutamate). Although a variety of calcium conductances exist, the resulting calcium current is depolarization sensitive and always inward owing to the vast difference between the intracellular (μ moles) and extracellular (m moles) calcium concentrations that barely generate any outward current beyond the reversal potential; if inactivation exists, it is much slower than that for sodium currents. Three prominent types of calcium currents, namely, high threshold Ca^{2+} current (I_{CaL}), low-threshold Ca^{2+} current (I_{CaT}), and intermediate threshold Ca^{2+} current (I_{CaN}) have been modeled as

$$\begin{aligned}
I_{\mathrm{CaL}} &= m(V_{\mathrm{m}})^k h([\mathrm{Ca}^{2+}]_i) I_{\mathrm{Ca}},\\
I_{\mathrm{CaT}} &= m(V_{\mathrm{m}})^2 h(V_{\mathrm{m}}) I_{\mathrm{Ca}},\\
I_{\mathrm{CaN}} &= m(V_{\mathrm{m}})^3 h(V_{\mathrm{m}}) I_{\mathrm{Ca}},
\end{aligned} \tag{2.11}$$

where k is an integer ≥ 2, and m and h are activation and inactivation functions of the calcium currents, respectively. Note that m is voltage dependent, while h can be calcium (for I_{CaL}) or voltage dependent (for $I_{\mathrm{CaT}}$and I_{CaN}). The term I_{Ca} can be given by the Goldman–Hodgkin–Katz current equation or its variants, or by Ohm's law (see Ref. [19], Chapter 9).

Several types of potassium currents that depend on the intracellular calcium concentration exist and are known as calcium-dependent potassium (I_{KCa}) currents. These currents primarily serve to limit firing frequencies and/or cause firing frequency adaptation. Two main I_{KCa} currents include the voltage and calcium-dependent big conductance potassium currents (I_{BK}) and calcium-dependent small conductance potassium currents (I_{SK}). These currents can be modeled as

$$\begin{aligned}
I_{\mathrm{BK}} &= g_{\mathrm{BK}} m(V_{\mathrm{m}}, [\mathrm{Ca}^{2+}]_i)(V_{\mathrm{m}} - E_{\mathrm{K}}),\\
I_{\mathrm{SK}} &= g_{\mathrm{SK}} m([\mathrm{Ca}^{2+}]_i)(V_{\mathrm{m}} - E_{\mathrm{K}}).
\end{aligned} \tag{2.12}$$

The I_{SK} current is typically referred to as I_{AHP} and is responsible for a second component of the after-hyperpolarization following a spike.

2.3.5
Simplified Neuronal Models

Large-scale network models based on detailed structural and physiological components such as outlined in the previous sections result in increased computational burden. Hence, in order to ask functional questions on the role of individual neurons in encoding information, simplified models of neurons have often been used. Moreover, simplified models can guide the design and development of biologically based neural networks. These models are based on the premise that all information in the nervous system is essentially encoded by the firing rates (*rate code*) of cells or by temporal patterns (*temporal code*) of spikes (also see Ref. [19], Chapters 14 and 15). The two most common types of models are *spike/pulse* models and *firing rate* models.

A well-known example for spike models is the *integrate-and-fire* (IF) model that is essentially a voltage integrator and generates impulses/spikes when a certain voltage threshold (V_{th}) is reached. IF models existed long before the physiological description of mechanisms of action potential and the HH model. The original investigation of a basic form of IF model was performed by Lapique [35, 36]; other examples include [37, 38]. The general description of the membrane current balance equation for the IF model is given by

$$C\frac{dV(t)}{dt} = -\frac{V(t)}{R} + I(t). \tag{2.13}$$

Upon crossing the threshold V_{th}, discrete, all-or-none output impulses are generated that can be represented by a series of delta functions $\sum_i \delta(t-t_i)$; t_i is the time at which an ith spike occurs, where $i = 1, 2 \cdots n$. The continuous firing rate as a function of input current I for the above case is given by $F = 1/T_{th}$, where T_{th} is the time to spike for a current $I > I_{th}$, given by $T_{th} = -\tau \ln(1-(V_{th}/IR))$; the time constant $\tau = RC$. Here I_{th} is the threshold current required for spike generation.

Several variants of the IF model exist. For example, the one described above is a *leaky* IF model due to the existence of the resistive component of the current. Alternatively, *nonleaky* IF models lack the resistance R and hence include only the capacitive current.

There are also certain forms of reduced models that can generate spikes in a threshold-dependent manner similar to IF models while retaining some of the physiological characteristics of the HH type models in reduced form, providing an approach to improve efficacy of large-scale computations (e.g., [39]). Moreover, such models are mathematically more tractable compared to their empirically realistic counterparts [40]. On the other hand, IF models are good candidates for electronic circuit implementation of neuromorphic designs using circuit elements such as capacitors, resistors, current/voltage sources, and switching components (see Chapters 4 and 6).

Firing rate models define firing frequency as a continuous function. Experimentally, the firing rates are obtained using three different procedures including averaging over time, averaging over several runs, or averaging across a population. When modeling a firing rate neuron, the rate is typically modeled as a continuous function of membrane voltage (e.g., $F = g(V)$, where $g(V)$ is often a sigmoidal function [19]). Alternatively, considering averaging over time, the average firing rate F can be expressed as a function of the input (intracellular current) I, which remains constant over a period $T\,(F = g(I))$. Extending this to a network of neurons, the input I_i to a postsynaptic neuron i is generated by the rates F_j of other (presynaptic) neurons j given by

$$F_i = g\left(\sum_j w_{ij} F_j\right), \qquad (2.14)$$

where w_{ij} is a weighting factor representing synaptic efficacy.

Choices for the gain function $g(\cdot)$ include logistic function [22] or hyperbolic tangent [17]. We further refer interested readers to the highly influential Wilson–Cowan equations [41] in the neural network community, describing the dynamics of interaction between simple excitatory and inhibitory populations of neurons and also to the derivations of biophysically motivated heuristic and conductance-based networks considered by Ermentrout and Terman [22]. In summary, firing rate models have greatly influenced artificial neural networks such as multilayer perceptrons and massively recurrent attractor networks as well as models of biological phenomena.

2.4
Synaptic Coupling and Plasticity

The well-known pioneer of neuroscience, Santiago Ramón Y Cajal, first proposed that memories might be formed by strengthening the connections between existing neurons [42]. Synaptic plasticity was more formally proposed as an underlying mechanism for learning and memory in the theoretical work of Hebb, who postulated that a connection between two neurons is potentiated when the spiking activity of one neuron drives the spiking activity of the other [9]. Later theoretical work suggested that decreases in synaptic strength are also necessary for learning in order to prevent the saturation of synaptic strengths and retain selectivity [43].

Many neural correlates exist for these theoretical ideas. Paired pulse facilitation and depression are well known forms of short-term synaptic plasticity, where a change in the amplitude of an excitatory postsynaptic conductance (EPSC) is observed when the eliciting presynaptic spike follows an earlier conditioning spike. These and similar forms of short-term changes to chemical synapses, which last for at most a few minutes, are often attributed to the effects of a residual elevation in presynaptic Ca^{2+} [44]. In contrast, long-term potentiation (LTP) and long-term depression (LTD) are persistent, lasting from several minutes to hours or longer. LTP refers to a long-lasting, activity-dependent enhancement in synaptic efficacy, whereas LTD is a persistent, activity-dependent reduction in the efficacy of chemical synapses. Both LTP and LTD occur due to multiple biophysical mechanisms, which vary across brain regions, species, and development stages [45, 46]. Spike timing dependent plasticity (STDP) is a form of Hebbian learning that relies on tight temporal correlations between the spikes of presynaptic and the spikes of postsynaptic neurons. With STDP, when presynaptic action potentials repeatedly arrive a few milliseconds before the generation of postsynaptic action potentials, long-term potentiation of the synapse occurs. However, the repeated occurrence of postsynaptic action potentials just prior to the arrival of presynaptic action potentials leads to long-term depression. It is widely believed that STDP is critically important both for learning and memory and for the development of the nervous system [47, 48].

2.4.1 Modeling Synaptic Plasticity

A popular phenomenological model for short-term plasticity comes from Tsodyks and Markram [49, 50], where model synapses can be depressing, facilitating, or have both depressing and facilitating components. This model assumes that a synapse is characterized by the residual synaptic resources following a number of synaptic events. Each presynaptic spike arriving at time t_{sp} activates a fraction of resources denoted by U. The corresponding kinetics can be written as

$$\begin{aligned}\frac{dx}{dt} &= \frac{z}{\tau_R} - Ux\delta(t-t_{sp}),\\ \frac{dy}{dt} &= -\frac{y}{\tau_i} + Ux\delta(t-t_{sp}),\\ \frac{dz}{dt} &= \frac{y}{\tau_i} - \frac{z}{\tau_R},\end{aligned} \tag{2.15}$$

where x, y, and z are the fractions of the resources in the recovered, active, and inactive states, respectively. The postsynaptic current is taken to be proportional to the fraction of resources in the active state so that $I_{syn}(t) = Ay(t)$, where the parameter A is the absolute synaptic strength seen when all resources are activated. Between synaptic events, replenishment of synaptic resources occurs based on linear dynamics with recovery time constant τ_R.

To include facilitation, we assume that U is not fixed but instead is increased with each presynaptic event. In this case, the additional equation for U is

$$\frac{dU}{dt} = -\frac{U}{\tau_F} + U_{SE}(1-U)\delta(t-t_{sp}), \tag{2.16}$$

where U_{SE} determines the increase in the value of U with each spike. This equation can be represented by a recursive expression for the value of U upon the arrival of $n+1$ action potentials,

$$U_{(n+1)} = U_n \exp\left(-\frac{\Delta t}{\tau_F}\right) + U_1\left(1 - U_n \exp\left(-\frac{\Delta t}{\tau_F}\right)\right), \tag{2.17}$$

where Δt is the duration of the period between spike number n and spike number $n+1$, and τ_F is the facilitation time constant. Note that between synaptic events, there is a linear decay of this accumulated fraction of resources that will be employed by the next spike with a single time constant τ_F. In the limit as $\tau_F \to 0$, the synapse is depressing only since no facilitation can accumulate. In the limit as $\tau_R \to 0$, the synapse is facilitating only since all resources are instantly replenished.

The discovery of spike timing dependent plasticity has led to the development of many models where synapses are potentiated or depressed based on the relative timing between each pair of presynaptic and postsynaptic action potentials. In the widely used STDP model of Song *et al.* [51], synaptic modification is defined by the learning rule,

$$F(\Delta t) = A_+ \exp\left(\frac{\Delta t}{\tau_+}\right) \tag{2.18}$$

for spike pairs with negative timing difference $\Delta t = t_{pre} - t_{post}$, where the presynaptic spike occurs before the postsynaptic spike. In contrast,

$$F(\Delta t) = -A_- \exp\left(-\frac{\Delta t}{\tau_-}\right) \tag{2.19}$$

for spike pairs with positive timing difference, where the postsynaptic spike occurs before the presynaptic spike. According to this learning rule, $\bar{g}_{syn}$ is replaced by $\bar{g}_{syn} + F(\Delta t)\bar{g}_{max}$. That is, the parameters A_+ and A_- control the maximum size of the update to the peak synaptic conductance as a fraction of the maximum possible synaptic conductance $\bar{g}_{max}$ for potentiation and depression, respectively. The time constants τ_+ and τ_- determine the time differences for which positive and negative timing differences lead to modification of the synaptic strength, respectively. Generally, this STDP model truncates changes to the synaptic strength that would lead to weights outside of a range between zero and the upper bound $\bar{g}_{max}$.

In contrast to these phenomenological models, there are many biophysically based models of LTD, LTP, and STDP that concentrate on modeling the biochemical and physiological processes that lead to synaptic changes. Most of these models focus on how synaptic plasticity depends on the dynamics of calcium. In particular, many models rely on the calcium control hypothesis that postulates that a large calcium transient results in LTP, while a moderate increase results in LTD [52–54]. Other more detailed biophysically based models explicitly model the signal transduction pathways associated with synaptic plasticity [53, 55, 56].

2.5
Computational Models of Neuronal Systems for Biohybrid Applications

Design of biohybrid systems requires suitable interfaces that enable robust communication between the biological and the nonbiological systems. The interface design often refers to the electrodes and tissue–hardware interactions. As discussed in Chapter 1, the whole system, however, requires the ability to decode and process the information from the living system or influence that activity of the living system. Computational models of neurons and neuronal networks can support the design of the biohybrid systems in several ways. Software implementations of computational models can be integrated as part of the living neuronal network in real time. Chapters 5 and 7 describe the dynamic clamp approach, a hybrid interfacing technique that allows direct interaction of physiologically realistic models with biological neuronal networks *in vitro*. Computational neural models or finite element modeling of the neural structures themselves can also be used in the design of neurotechnology to stimulate the nervous system [57–62]. Chapter 8 describes the use of biologically inspired neural models in the design of systems to control non-neuronal components that interact with humans for rehabilitation therapy and other activities. They can also be used in the design of closed loop biohybrid systems for control of physiological function with wide ranging applications such as neurocardiology (Chapter 10). Simple conductance-based and cable theory-based neuronal models provide the theoretical basis for designing electronic neuromorphic circuit design [63–65] that can be directly interfaced with the nervous system [66]. Chapter 3 presents basic principles for design of neuromorphic hardware, while Chapters 9 and 11 illustrate the use of neuromorphic hardware for various applications.

2.6
Resources

1) **NEURON**: Simulation tool for empirically based models of neurons and networks. Web site: http://www.neuron.yale.edu/neuron/; book: [67].
2) **GENESIS**: The General Neural Simulation Software, a simulation tool for realistic modeling of subcellular components and biochemical reactions to single neurons, networks, and system-level models. Web site: http://genesis-sim.org/; book: [6].
3) **XPPAUT**: A general numerical simulation tool for simulating, animating, and analyzing dynamical systems. It is widely used by computational and theoretical neuroscientists as it provides the ability to perform a complete numerical analysis of the dependence of solutions on parameters (bifurcation analyses). Web site: http://www.math.pitt.edu/~bard/xpp/xpp.html; book: [16].
4) Some of the other available tools are reviewed in Ref. [68].

Acknowledgments

This work was partially supported by a US National Institutes of Health grant (R01-NS054282) to R.J.

References

1 Hodgkin, A.L. and Huxley, A.F. (1952) A quantitative description of membrane current and its application to conduction and excitation in nerve. *J. Physiol.*, **117** (4), 500–544.

2 Rall, W. (1957) Membrane time constant of motoneurons. *Science*, **126** (3271), 454.

3 Rall, W. (1959) Branching dendritic trees and motoneuron membrane resistivity. *Exp. Neurol.*, **1**, 491–527.

4 Rall, W. (1960) Membrane potential transients and membrane time constant of motoneurons. *Exp. Neurol.*, **2**, 503–532.

5 Rall, W. (1964) Theoretical significance of dendritic trees for neuronal input–output relations, in *Neural Theory and Modeling* (ed. R.F. Reiss), Stanford University Press.

6 Bower, J.M. and Beeman, D. (1998) *The Book of GENESIS: Exploring Realistic Neural Models with the General Neural Simulation System*, 2nd edn, Springer, New York.

7 Hines, M.L. and Carnevale, N.T. (1997) The NEURON simulation environment. *Neural Comput.*, **9** (6), 1179–1209.

8 Schwartz, E.L. (ed.) (1990) *Computational Neuroscience*, MIT Press.

9 Hebb, D.O. (1949) *The Organization of Behavior: A Neuropsychological Theory*, John Wiley & Sons, Inc., New York.

10 Hille, B. (1992) *Ionic Channels of Excitable Membranes*, 2nd edn, Sinauer Associates.

11 Johnston, D. and Wu, S.M. (1995) *Foundations of Cellular Neurophysiology*, MIT Press, Cambridge, MA.

12 Kandel, E.R. Schwartz, J.H., and Jessell, T.M. (eds) (2000) *Principles of Neural Science*, 4th edn, McGraw-Hill.

13 Schmidt, R.F. (ed.) (1978) *Fundamentals of Neurophysiology*, 2nd edn, Springer, New York, p. 339.

14 Shepherd, G.M. (1994) *Neurobiology*, 3rd edn, Oxford University Press, New York.

15 Zigmond, M.J. and Squire, L.R. (eds) (1999) *Fundamental Neuroscience*, Academic Press, San Diego, CA.

16 Ermentrout, B. (2002) *Simulating, Analyzing and Animating Dynamical Systems*, Society for Industrial and Applied Mathematics, Philadelphia.

17 Gerstner, W. and Kistler, W. (2002) *Spiking Neuron Models: Single Neurons, Populations, Plasticity*, Cambridge University Press, Cambridge, UK.

18 Keener, J. and Sneyd, J. (2009) Mathematical physiology, in *Interdisciplinary Applied Mathematics*, 2nd edn (eds S.S. Antman, J.E. Marsden, and L. Sirovich), Springer, New York.

19 Koch, C. (1999) *Biophysics of Computation* (ed. M. Stryker), Oxford University Press, New York.

20 Koch, C. and Segev, I. (eds) (1998) *Methods in Neuronal Modeling: From Ions to Networks*, 2nd edn, MIT Press.

21 Rall, W. (1977) Core conductor theory and cable properties of neurons, in *Handbook of Physiology, Cellular Biology of Neurons* (eds E.R. Kandel, J.M. Brookhardt, and V.M. Mountcastle), American Physiological Society, Bethesda, MD, pp. 39–97.

22 Ermentrout, G.B. and Terman, D.H. (2010) *Foundations of Mathematical Neuroscience*, Springer, New York.

23 Finger, S. (2001) *Origins of Neuroscience: A History of Explorations into Brain Function*, Oxford University Press, USA.

24 Harris-Warrick, R.M. (2002) Voltage-sensitive ion channels in rhythmic motor systems. *Curr. Opin. Neurobiol.*, **12** (6), 646–651.

25 Hodgkin, A.L. and Huxley, A.F. (1952) The components of membrane conductance in

the giant axon of Loligo. *J. Physiol.*, **116** (4), 473–496.

26 Hodgkin, A.L. and Huxley, A.F. (1952) Currents carried by sodium and potassium ions through the membrane of the giant axon of Loligo. *J. Physiol.*, **116** (4), 449–472.

27 Hodgkin, A.L. and Huxley, A.F. (1952) Movement of sodium and potassium ions during nervous activity. *Cold Spring Harb. Symp. Quant. Biol.*, **17**, 43–52.

28 Hodgkin, A.L., Huxley, A.F., and Katz, B. (1952) Measurement of current–voltage relations in the membrane of the giant axon of Loligo. *J. Physiol.*, **116** (4), 424–448.

29 Fall, C.P. *et al.*. (eds) (2000) *Computational Cell Biology*, vol. **20**, Springer, New York.

30 Rinzel, J. (1985) Excitation dynamics: insights from simplified membrane models. *Fed. Proc.*, **44** (15), 2944–2946.

31 Fitzhugh, R. (1960) Thresholds and plateaus in the Hodgkin–Huxley nerve equations. *J. Gen. Physiol.*, **43**, 867–896.

32 Fitzhugh, R. (1961) Impulses and physiological states in theoretical models of nerve membrane. *Biophys. J.*, **1** (6), 445–466.

33 Hausser, M., Spruston, N., and Stuart, G.J. (2000) Diversity and dynamics of dendritic signaling. *Science*, **290** (5492), 739–744.

34 Mel, B.W. (1993) Synaptic integration in an excitable dendritic tree. *J. Neurophysiol.*, **70** (3), 1086–1101.

35 Brunel, N. and van Rossum, M.C. (2007) Lapicque's 1907 paper: from frogs to integrate-and-fire. *Biol. Cybern.*, **97** (5–6), 337–339.

36 Lapicque, L. (1907) Recherches quantitatifs sur l'excitation electrique des nerfs traitée comme une polarisation. *J. Phys. (Paris)*, **9**, 620–635.

37 Stein, R.B. (1967) The frequency of nerve action potentials generated by applied currents. *Proc. R. Soc. Lond. B Biol. Sci.*, **167** (6), 64–86.

38 Stein, R.B. (1967) Some models of neuronal variability. *Biophys. J.*, **7** (1), 37–68.

39 Izhikevich, E.M. (2003) Simple model of spiking neurons. *IEEE Trans. Neural. Netw.*, **14** (6), 1569–1572.

40 Izhikevich, E.M. (2007) *Dynamical Systems in Neuroscience: The Geometry of Excitability and Bursting*, The MIT Press, Cambridge, MA.

41 Wilson, H.R. and Cowan, J.D. (1972) Excitatory and inhibitory interactions in localized populations of model neurons. *Biophys. J.*, **12** (1), 1–24.

42 Ramon y Cajal, S. (1894) The Croonian Lecture: La fine structure des centres nerveaux. Proceedings of the Royal Society of London, London.

43 Sejnowski, T.J. (1977) Statistical constraints on synaptic plasticity. *J. Theor. Biol.*, **69** (2), 385–389.

44 Zucker, R.S. and Regehr, W.G. (2002) Short-term synaptic plasticity. *Annu. Rev. Physiol.*, **64**, 355–405.

45 Malenka, R.C. and Bear, M.F. (2004) LTP and LTD: an embarrassment of riches. *Neuron*, **44** (1), 5–21.

46 Massey, P.V. and Bashir, Z.I. (2007) Long-term depression: multiple forms and implications for brain function. *Trends Neurosci.*, **30** (4), 176–184.

47 Bi, G. and Poo, M. (2001) Synaptic modification by correlated activity: Hebb's postulate revisited. *Annu. Rev. Neurosci.*, **24**, 139–166.

48 Sjostrom, P.J. *et al.* (2008) Dendritic excitability and synaptic plasticity. *Physiol. Rev.*, **88** (2), 769–840.

49 Tsodyks, M.V. and Markram, H. (1997) The neural code between neocortical pyramidal neurons depends on neurotransmitter release probability. *Proc. Natl. Acad. Sci. USA*, **94** (2), 719–723.

50 Tsodyks, M., Uziel, A., and Markram, H. (2000) Synchrony generation in recurrent networks with frequency-dependent synapses. *J. Neurosci.*, **20** (1), RC50.

51 Song, S., Miller, K.D., and Abbott, L.F. (2000) Competitive Hebbian learning through spike-timing-dependent synaptic plasticity. *Nat. Neurosci.*, **3** (9), 919–926.

52 Karmarkar, U.R. and Buonomano, D.V. (2002) A model of spike-timing dependent plasticity: one or two coincidence detectors? *J. Neurophysiol.*, **88** (1), 507–513.

53 Lisman, J. (1989) A mechanism for the Hebb and the anti-Hebb processes underlying learning and memory. *Proc. Natl. Acad. Sci. USA*, **86** (23), 9574–9578.

54 Shouval, H.Z., Bear, M.F., and Cooper, L.N. (2002) A unified model of NMDA receptor-dependent bidirectional synaptic plasticity. *Proc. Natl. Acad. Sci. USA*, **99** (16), 10831–10836.

55 Ajay, S.M. and Bhalla, U.S. (2004) A role for ERKII in synaptic pattern selectivity on the time-scale of minutes. *Eur. J. Neurosci.*, **20** (10), 2671–2680.

56 Castellani, G.C. *et al.* (2005) A model of bidirectional synaptic plasticity: from signaling network to channel conductance. *Learn Mem.*, **12** (4), 423–432.

57 Grinberg, Y. *et al.* (2008) Fascicular perineurium thickness, size, and position affect model predictions of neural excitation. *IEEE Trans. Neural. Syst. Rehabil. Eng.*, **16** (6), 572–581.

58 Foutz, T.J. and McIntyre, C.C. (2010) Evaluation of novel stimulus waveforms for deep brain stimulation. *J. Neural. Eng.*, **7** (6), 066008.

59 Frankemolle, A.M. *et al.* (2010) Reversing cognitive-motor impairments in Parkinson's disease patients using a computational modelling approach to deep brain stimulation programming. *Brain*, **133**Pt (3), 746–761.

60 Stites, E. and Abbas, J.J. (2000) Sensitivity and versatility of an adaptive system for controlling cyclic movements using functional neuromuscular stimulation. *IEEE Trans. Biomed. Eng.*, **47** (9), 1287–1292.

61 Woock, J.P., Yoo, P.B., and Grill, W.M. (2010) Finite element modeling and *in vivo* analysis of electrode configurations for selective stimulation of pudendal afferent fibers. *BMC Urol.*, **10**, 11.

62 Holdefer, R.N., Sadleir, R., and Russell, M.J. (2006) Predicted current densities in the brain during transcranial electrical stimulation. *Clin. Neurophysiol.*, **117** (6), 1388–1397.

63 Cymbalyuk, G. *et al.* (2000) Modeling alternation to synchrony with inhibitory coupling: a neuromorphic VLSI approach. *Neural Comput.*, **12** (10), 2259–2278.

64 Le Masson, S. *et al.* (1999) Analog circuits for modeling biological neural networks: design and application. *IEEE Trans. Biomed. Eng.*, **46** (6), 638–645.

65 Bartolozzi, C. and Indiveri, G. (2007) Synaptic dynamics in analog VLSI. *Neural Comput.*, **19** (10), 2581–2603.

66 Jung, R., Brauer, E.J., and Abbas, J.J. (2001) Real-time interaction between a neuromorphic electronic circuit and the spinal cord. *IEEE Trans. Neural Syst. Rehab. Eng.*, **9** (3), 319–326.

67 Carnevale, N.T. and Hines, M.L. (2006) *The NEURON Book*, Cambridge University Press, Cambridge.

68 Brette, R. *et al.* (2007) Simulation of networks of spiking neurons: a review of tools and strategies. *J. Comput. Neurosci.*, **23** (3), 349–398.

3
Neuromorphic Electronic Design

Francesco Tenore and Ralph Etienne-Cummings

At the foundation of complex neuromorphic systems are simple, fundamental circuits that allow engineers to emulate the sophisticated behaviors of biological processing systems. To identify these circuits, it is first necessary to properly understand the properties of the underlying biological systems that they morph. The task then becomes to identify device properties or circuit characteristics that behave similarly in response to similar inputs. This chapter provides an overview of the fundamental devices and circuits that allow emulation of neuronal behavior as well as more complex sensory system processing circuits, to emulate the visual and auditory systems. For each processing element, applications of the circuits are shown. Finally, it provides models of circuits that can act on the environment, such as repetitive neural control patterns and muscle actuation.

3.1
Choices for Neuromorphic Circuits: Digital versus Analog

Think of an animal. Better yet, think of an insect. Every task that your insect specializes in doing is done better and many times more efficiently than any man-made machine attempting to perform similar tasks. For example, the most sophisticated aerial vehicle is no match for a bee's flight motor skills, even when a man is remote controlling it! The bee's rudimentary sensors allow it to scan large areas to find suitable locations for beehives and flowers for pollination. Comparing this "primitive" life form to state-of-the-art autonomous navigational robotic vehicles such as those participating in the Defense Advanced Research Projects Agency's Grand Challenge, which relied heavily on precise Global Positioning System sensor data using approximately 1 kW of computing power [1], or to the Air Force's Unmanned Aerial Vehicles that require remote guidance for navigation shows how distant our technology is to emulating even the most rudimentary life forms.

It is, in turn, possible to harness the power of neuronal computations using supercomputers, which rely on fast, clocked, two-dimensional multiprocessor digital systems characterized by high-power dissipation and large scales (when compared to ~1 cm^3

Biohybrid Systems: Nerves, Interfaces, and Machines, First Edition. Edited by Ranu Jung.

total volume occupied by an average bee). In contrast, however, neuronal activity is a slow, three-dimensional asynchronous process, requiring comparatively little energy. This discrepancy implies that although it may be possible to mimic neural behavior using these digital machines, they are not direct translations of neuronal circuits.

The analog nature of neuronal computations, as shown in the next sections, is such that the proper emulation of neuronal behavior requires devices and/or circuits that under appropriate conditions behave as neurons. The field of neuromorphic engineering stems from this premise.

Nonetheless, modern forms of neuromorphic engineering take advantage of many of the benefits of digital circuits. For example, the communications between analog or digital neurons use a protocol called Address Event Representation (AER), in which asynchronous digital circuits are used to multiplex multiple neuronal spikes onto a common data bus [2]. Furthermore, there is a movement toward using Field Programmable Gate Arrays (FPGA) to model large networks of neurons with digital hardware because of the reprogrammability and "instant" prototyping capabilities [3]. Hence, the debate between the purists – who hold to the original "Mead"-ian proposition of morphing nervous system using subthreshold analog integrated circuits – and the modern perspective – which uses any tool that allows the closest mimicry of biological performance – rages on. In this chapter, we will present examples from both camps.

Fundamentally, the circuits that are used in neuromorphic engineering are no different from what is used in standard analog and digital integrated circuit design. Hence, the basic building blocks, which can be found in a number of dedicated texts, will not be reviewed here. See references [4–6] for details on basic circuit design. On the other hand, we present the circuits as a neuromorphic engineer would design them. That is, we identify the biological system being implemented and then identify the circuits that can be used to realize the desired functionality. In this way, we restate some basic principles found in other chapters dedicated to the biology, but then translate the biophysical models into their circuit counterparts. One additional word of caution should be given here: this transformation is *not* unique. Hence, the examples provided are either the seminal contributions to the field or particular versions favored by the authors.

3.2
The Breadth of Neuromorphic Systems

The term "neuromorphic" was coined by Carver Mead in the late 1980s to describe the design of artificial systems that emulate the physical properties or information representations found in biological nervous systems [4].

The neuromorphic engineering approach to solve real-world problems typically stems from three important considerations:

- a biological entity (such as a cell, an organ, or an organ system) exists that solves a problem efficiently and elegantly;

- the biological entity is the fruit of billions of years of evolution, resulting in its present form;
- understanding how the problem is solved and why it is solved in that particular way allows the possibility of speeding up the evolutionary process for "artificial" mimics of the biological system, thereby providing alternative means to solve engineering problems.

In addition, it is also believed that the process of analyzing, building, and testing the neuromorphic systems can provide some insight into the construction and function of their biological counterparts. This perspective is fortified by the realization that the artificial version can be configured in modes that may be impossible for the biological version. To date this belief has not been fully verified, primarily because the all details of the biological systems are not fully understood, which makes their mimicry impossible. Nonetheless, the three motivations above, coupled with the final hope, have taken the field of neuromorphic engineering from single labs in the United States [4] and Switzerland [7] to tens of labs around the world.

As pointed out by one of the pioneers of the field, Misha Mahowald, perhaps one of the more fundamental differences between the biological systems and the silicon that seeks to emulate them is that the silicon systems are typically designed to perform a particular function, and its circuits are defined by their functional significance, rather than an arbitrary morphological demarcation [8]. Yet, neuromorphic engineers continue to morph silicon into models of biology with the hope of unlocking the computational magic of the latter. In the following sections, we show how this is done and some of the successes that have been achieved.

3.3 The Fundamental Processing Unit: The Neuron

To gain insight into the way the nervous system computes, we start with its fundamental unit of computational activity: the neuron. The previous chapter presented an overview of the physiology of neurons and general formalisms in neuronal modeling. Here we focus on understanding how these neural models can be implemented in basic electronic circuits.

Chapter 2 detailed the many different charge carriers (Na, K, Ca) involved in neural processes. These characteristics of the action potential can be reproduced (to some extent) using a single charge carrier (the electron) exploiting the physical laws responsible for the exponential relationship between current and voltage of an MOS transistor operating in the *subthreshold* or *weak inversion* region. A transistor is said to be operating in this region if its gate-to-source voltage is below the transistor's specified threshold voltage. In this case, the transistor's drain current, I_D, is calculated as

$$I_D = I_0 e^{\kappa V_{gs}} (1 - e^{-V_{ds}}), \quad (3.1)$$

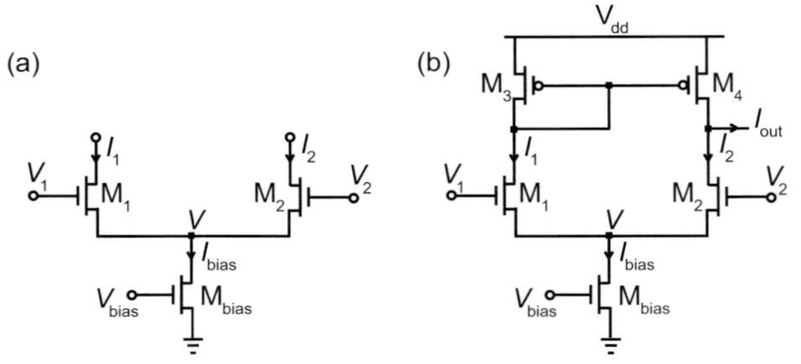

Figure 3.1 (a) Schematic of the differential pair. The bias voltage V_{bias} is set to ensure operation in the subthreshold region. (b) Schematic of an operational transconductance amplifier. The current I_{out} is proportional to the difference between V_1 and V_2.

where κ is a function of the geometry of the device and is inversely proportional to the thermal voltage, $U_t = kT/q$.

Similarly, the behavior of the cell membrane can be emulated by a fixed capacitor, and the voltage-dependent conductances of biological membranes in steady state can be emulated by the current–voltage relationship generated by CMOS transistors arranged to form a differential pair, as shown in Figure 3.1a [4]. In this circuit, V_{bias} is a subthreshold voltage ($V_{bias} < V_{th}$) and in normal operation its drain voltage, V, is saturated, making the $e^{-V_{ds}}$ term in Equation 3.1 negligible. The contributions to I_{bias} due to M_1 and M_2 are therefore a function of the difference between V_1 and V_2. Using Equation 3.1,

$$I_{bias} = I_1 + I_2 = I_0 e^{(V_1 - V)} + I_0 e^{(V_2 - V)} = I_0 e^{V}(e^{V_1} + e^{V_2}) \tag{3.2}$$

or

$$\begin{aligned} I_1 &= I_{bias}\frac{e^{V_1}}{e^{V_1} + e^{V_2}}, \\ I_2 &= I_{bias}\frac{e^{V_2}}{e^{V_1} + e^{V_2}}. \end{aligned} \tag{3.3}$$

Mirroring I_1 onto I_2 allows computation of the difference between the two currents, as shown in Figure 3.1b. This circuit is an *operational transconductance amplifier* (OTA) and is characterized by the fact that its output current, I_{out}, is equal to the difference between I_1 and I_2 and is therefore a function of the difference between V_1 and V_2:

$$\Delta I = I_{out} = I_1 = I_2 = I_b \tanh\left(\frac{\kappa(V_1 - V_2)}{2}\right) \tag{3.4}$$

and if $V_1 \approx V_2$, then $\tanh(\Delta V) \approx \Delta V$,

$$G_m = \frac{\partial I_{out}}{\partial V_{in}} \approx \frac{\kappa I_b}{2}. \tag{3.5}$$

Recall from Equation 3.1, that as κ is inversely proportional to the thermal voltage U_T, the κI_b term in Equation 3.5 is, in fact, a conductance. Also note that the maximum current for each conductance is directly controlled by its V_{bias}.

The activation and inactivation of membrane conductances can then be emulated by the voltage-dependent output currents from differential pairs. The ability to control this conductance allows the implementation of $I(V) = g(E - V)$, as discussed in Equation 2.1.

Starting from this foundation, we proceed to examine in more detail how silicon neurons can be designed, in relation to the type of functionality that they seek to emulate. At the two ends of the "neuron emulation" spectrum, we find neurons that seek to faithfully capture the dynamics of neural signals, thereby implementing circuits that qualitatively follow Hodgkin–Huxley or Morris–Lecar models [9], and at the other end are circuits that seek to capture the neurons' basic functionality focusing instead on the massive interconnectivity that broadly characterizes neural systems. This is typically achieved by using simpler models such as integrate-and-fire models, described in Section 3.3.2.2 [4]. The implicit trade-off involved in the two different approaches is between the fidelity of the dynamics of the circuits that are emulated and the area required for the circuits (or complexity of computation). Also see Refs [10, 11] for an in-depth review of silicon neurons.

3.3.1 Conductance-Based Modeling

The first examples of analog integrated circuits with dynamics similar to those of real neurons, as dictated by the Hodgkin–Huxley model, were engineered in the late 1980s by Mahowald and Douglas [12]. The circuits emulated ion currents in neurons in the neocortex, in which the activation and inactivation of the membrane conductances controlling the flow of sodium-like and potassium-like ions depended on both time and voltage.

Figure 3.2 helps illustrate how these types of circuits can be constructed. Start by assuming a time-varying, nonzero voltage V_{mem}. In this case, O_{Na} and O_K, which are OTAs, are configured as buffers and low-pass filters for the input voltage. The sodium and potassium conductance dynamics can then be tuned through $V\tau_{Na}$ and $V\tau_K$. The conductance-tuning voltages V_{Na_Off}, V_{Na_On}, and V_K produce inactivation and activation sodium currents and potassium currents, respectively, which contribute to the charging and discharging of the membrane potential. Finally, the threshold voltage is a global parameter that is also responsible for modulating the charging and discharging currents.

Additional elements that can be added to this basic structure with similar modules are a *leakage* current, a *persistent (bias) sodium* current, a *calcium-dependent potassium current* (to add adaptation), and so on.

Similarly, Patel and DeWeerth [13] and Simoni *et al.* [14] implemented a Morris–Lecar model of a neuron in their silicon implementation. These models are characterized by the presence of a slow time constant in conjunction with the action potential-generating fast time constant, typically found in high-fidelity neuron modeling.

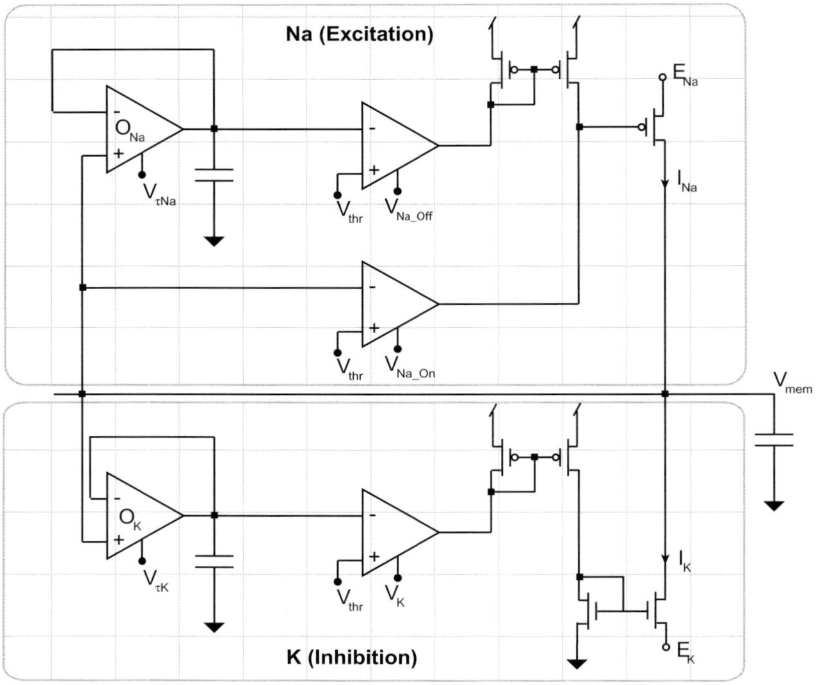

Figure 3.2 Schematic of conductance-based model for reproducing the dynamics of sodium (excitatory) and potassium (inhibitory) currents.

It should be noted that regardless of the type of neuron modeled, or the way a neuron is modeled, *no two neurons are ever exactly identical* in their characteristics. This is another feature that silicon neurons share with their biological counterparts. These differences are – in both cases, albeit differently – caused by "manufacturing," that is, nature or the integrated circuit processing.

3.3.2
Compartmental Modeling

With reference to Figure 2.1, neural signals can also be reproduced by individually emulating the functionality of the three compartments that give neural signals their characteristics: the somatic compartment, the metabolic center of the neuron, responsible for producing the action potentials; the dendritic compartment, which allows the cell body to connect to neighboring nerve cells and which "contains" the synapses that provide inputs to a neuron; and the axonal compartment, which sends a neuron's signal to adjacent neurons' synaptic inputs. The remainder of this section illustrates the inner workings of neuromorphic circuits that emulate each compartment.

3.3.2.1 The Dendritic Compartment: Home to the Synapses

As seen in the previous chapter, when an action potential traveling along a neuron's axon reaches a synaptic terminal, it provokes the release of neurotransmitting molecules from the terminal into a *synaptic cleft*. This is where the neurotransmitters bind themselves to receptor molecules on an adjacent (postsynaptic) neuron's dendrites, as captured in Figure 2.1 (box), causing ion channels to open, thus altering the membrane potential.

This alteration is at the root of one of the most important processes in neural systems: synaptic weighing or modulation, which gives rise to these systems' ability to "learn." Here, "learning" is viewed as the ability to use past experiences to appropriately modify the system in response to new stimuli.

In vertebrates, synapses are responsible for modulating these electrical membrane properties, and more in general the activity, of neurons. Each neuron, in fact, has thousands of synapses to receive information from and broadcast the integrated information to adjacent neurons.

In silicon, the synaptic modulation can be achieved mainly in two ways, again depending on specific implementation functions of the synapses. Specifically, simple synaptic models modulate incoming signals through externally programmable weighting schemes, thus making the learning circuitry external to the neuron. These synaptic voltages can be modulated using either externally programmable digital-to-analog voltages or externally controllable voltages for an operational transconductance amplifier. These circuits are typically configured off-chip and appropriate when fixed neural spiking rhythms are used, such as to emulate a central pattern generator (see Chapter 9 for details on these implementations).

The types of synapses described lack synaptic mechanisms of shunting inhibition, where the synaptic conductance short-circuits currents that are generated at adjacent excitatory synapses. This feature is typically used to reduce the effect of excitatory signals impinging on a neuron *after* the same neuron has fired an action potential. This inhibition of the excitatory postsynaptic potentials, or EPSPs, are at the foundation of the unsupervised learning mechanisms (short-term plasticity) of the brain, such as spike timing-dependent plasticity (STDP) introduced in Chapter 2. STDP depends on the causal relationship between the pre- and postsynaptic spikes to strengthen the synapse between two neurons (the anticausal relationship leads to synaptic depression) [15]. Shunting inhibition can be implemented compactly using a synaptic cleft-type circuit, capable of extending the duration of an action potential, and a low-pass filter [16, 17].

Silicon implementations of synapses that exhibit short-term plasticity are suitable for evaluating the computational roles of synaptic adaptation in large networks of spiking neurons using complex stimuli and in real time [18]. On the other hand, realization of long-term plasticity circuits, using transistors from the flash-memory technology [19], allows us to implement learning algorithms and set synaptic weights of synapses automatically, without requiring dedicated pins or wires for individual synapses.

3.3.2.2 The Somatic Compartment: Spike-Based Processing and the Integrate-and-Fire Model

English neurophysiologist and Nobel laureate Charles Sherrington considered the integrative action of the nervous system – producing its decision-making results – the most fundamental operation performed by the brain [20].

Integrate-and-fire neuron models, introduced in Chapter 2, are typically divided into the *nonleaky*, or ideal model, and leaky. At their core, these models are made up of a capacitor, acting as the integrator, a comparator, with the threshold voltage as trigger, and a discharging circuit.

Figure 3.3a shows a basic nonleaky integrate-and-fire model, and Figure 3.3b–h present some variations in the basic theme: in which the threshold voltage can be tuned (b); in which the leakage element can be added (c, modeled as a resistor that discharges the charge on the capacitor); where a tunable refractory period can be added (d); in which the output pulse width can be tuned, for example, through the use of current tuning bias voltage (e, V_{bpw}) or a hysteretic comparator (f, where V_p sets the difference between the comparator's high and low thresholds); where the resting

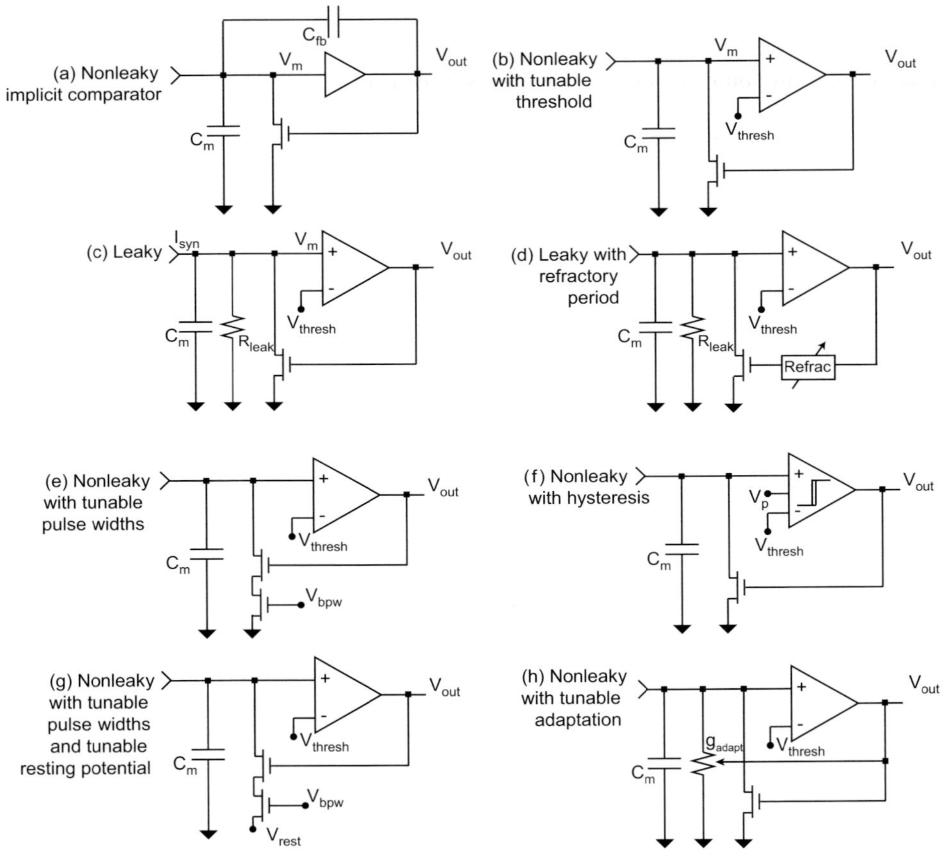

Figure 3.3 Basic integrate-and-fire (I-F) models.

potential can be tuned (g); and in which adaptation of the output spike frequency can be added to gradually remove charge from the capacitor, more spikes are generated by the neuron (also described as synaptic fatigue, h).

To illustrate the behavior of this model, we examine the circuit in Figure 3.3c and show how the results extend to other implementations.

Assuming an ideal comparator, the synaptic input current, I_{syn}, partly charges the membrane capacitor, C_m, and is partly dissipated on the resistance R_{leak}:

$$I_{syn} = C_m V_m + \frac{V_m}{R_{leak}}, \tag{3.6}$$

where V_m is the voltage across the membrane. Therefore, if a constant current $I_{syn} < V_{thresh}/R_{leak}$, then V_m will relax exponentially to the steady-state voltage ($I_{syn}R_{leak}$) but never reach the threshold voltage. In this situation, V_{out} is always low.

The current required to charge the membrane capacitor to the threshold voltage, such that $V_m = V_{thresh}$, is called the *rheobase*. If the input synaptic current is greater than the rheobase, then V_m reaches V_{thresh} and the comparator switches state. This causes the NMOS resetting transistor to turn on, thus removing all charge from C_m. The assumption made here is that the resetting is instantenous. In the model shown in Figure 3.3e, f and g, the pulse width is nonnegligible, and in (g) the capacitor discharges at most to V_{rest}. This causes the comparator to go low, and the cycle starts anew. The firing frequency is proportional to I_{syn} and significantly also depends on whether the model implements a nonzero pulse width (e,f), a refractory period (d), a leak (c,d), or spike frequency adaptation (SFA, h).

Typically, combinations of the many features that can be added to the simplest model are used. For example, Carver Mead's self-resetting neuron [4] is a combination of the nonleaky implicit comparator (Figure 3.3a) with tunable pulse width (Figure 3.3e). Indiveri *et al.* [21, 22] also use source followers to increase their linear integration range.

More recently, Folowosele *et al.* [23] reported on a different kind of integrate-and-fire neuron based on the Mihalas-Niebur model [24] and implemented using an adaptive threshold and a switched capacitor-based architecture to implement the model's parameters. Using the switched capacitor methodology instead of depending on the subthreshold transistor characteristics (as is typically the case in the neuromorphic engineering community) allows for better controllability, matching, and reproducibility of the neuron's behavior. These properties may not be important when there is redundancy through millions to billions of individual units (as in biology); however, it is important when there are limited neurons in the network and each neuron has to be as accurate as possible in representing the input signal, internal state, or in making decisions.

3.3.2.3 **The Axonal Compartment: Address-Event Representation**

Communication between neurons occurs mainly through the largest dendritic process, the axon. Experiments on a variety of sensory systems, such as locust olfaction, cat vision, and monkey vision and audition, suggest that spike timing

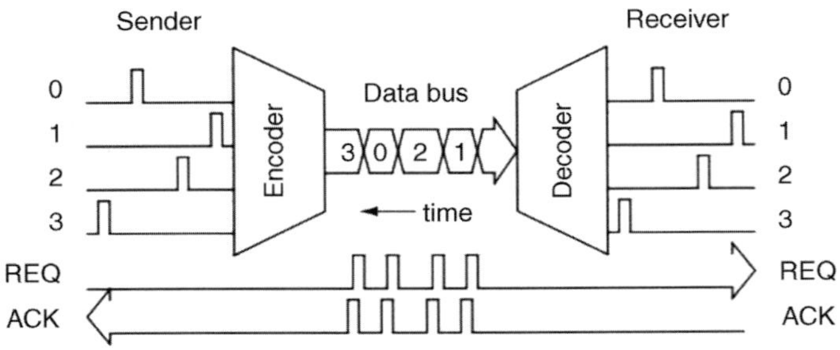

Figure 3.4 Address event representation (reproduced from Ref. [26]).

across neurons and neuron ensembles is crucial to understanding neuronal behavior in response to stimuli. Therefore, if we place our focus more on temporal coding, that is, the timing of the event, then it becomes clear that what we should seek is to acquire all spike occurrences. Address-Event Representation (AER) is used by the neuromorphic community to multiplex spikes from a population of neurons onto a common bus. This multiplexing takes advantage of the six orders of magnitude difference between the spike frequency of silicon neurons (when operating at physiological rates) and the inherent speed of silicon. Hence, each bus can carry the information from up to 1 million neurons. Because the spikes now share the common communication channel, it is important to know where they originated. Hence, the address of the source is attached to each spike or event, leading to the term Address-Events. If the bus request conflicts can be resolved quickly (i.e., in nanoseconds), then the timing of the spikes is preserved. This approach to spike communications is well described in Refs [2, 25]. This technique, of which a schematic representation is shown in Figure 3.4, allows neuromorphic systems to exploit the strengths of digital communication circuits, with their constantly increasing transmission speeds, to help create mappings between large numbers of neurons in two-dimensional space, thereby helping to offset the inherent three-dimensional advantage in biological neural systems. This allows complex neuronal architectures to be developed in silicon, as shown in the integrate-and-fire array transceivers developed by Goldberg *et al.* and Vogelstein *et al.* [26, 27], as well as in modeling of various areas of the brain, such as the visual cortex [28] and the hippocampus [17] and neuromorphically inspired image sensors [29].

3.4
Sensing the Environment

At a fundamental level, neural systems evolved to process sensory information required to survive in predatory environments. At the periphery of the nervous systems are complex structures that are capable of transducing environmental data in neural signals. In this section, we focus on the emulation of the transduction of light into neural signals, to emulate the visual system, and of sound waves into neural signals, thereby emulating the auditory system.

3.4.1
Vision

The human visual system is responsible for providing the brain with optical information, from which it can extract information relevant for its interaction with the environment. Given the difficulty of this task, the vision system is a complex, diverse structure that starts from the eyes' "mechanical" components, that is, the lens, iris, cornea, and pupil, and proceeds toward the neural transduction and preprocessing of the light information [30] to relay information to the brain's primary visual cortex (V1) and other areas of the brain. Among the most fascinating aspects of this structure is the transduction itself, which is accomplished by photosensitive neural tissue at the back of the eye, known as the retina, and its ability to quickly adapt to environmental changes in background illuminance, and to compressively code the information from 100 million photodetectors and communicate the information using only 1 million optic nerve fibers. Silicon implementation of such a complex system typically focuses on the neural front and back ends of the visual system. That is, the retina [29, 31–33] and the visual cortex [25–28] were implemented. This section will, therefore, examine the neuromorphic circuit emulation of these systems.

Neuromorphic vision starts with emulating one of the farthest outposts of the central nervous system: the retina. The retina is a thin sheet of neural, light-sensitive tissue that lines the inner surface of the eye. It is responsible for collecting all the visual information in the receptive field of view and provide initial processing for the brain to interpret. With reference to Figure 3.5, the retina's first layer is made

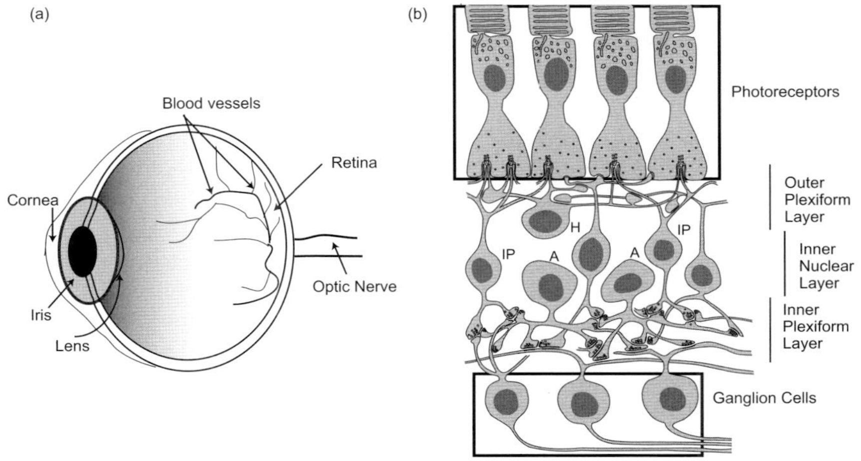

Figure 3.5 (a) Human eye; (b) cross section of primate retina, showing subset of all signal pathways. The photoreceptors transduce the light into neural signals, which are relayed in the outer plexiform layer. The intermediate bipolar and amacrine cells then relay the visual information to the ganglion cells, the axons of which form the optic nerve, which extends into the brain. A: amacrine cell; H: horizontal cell; IB: invaginating bipolar cell. Adapted from Dowling [30].

up of two types of photoreceptors, which amplify photon events. These photoreceptors are the rods and cones. Rods are active mainly under dim light conditions and provide black and white vision; cones support daytime vision and add the perception of color.

This initial transduction does not output discrete action potentials, the sort found in most other areas of the brain. In fact, all the neurons in the subsequent outer plexiform layer (OPL) and prior to the efferent synaptic connections to the ganglion cells, represent information with smooth gradient responses [30]. Specifically, with reference to Figure 3.5b, the cones make excitatory synapses onto both horizontal cells (H) and bipolar cells (IB) [30]. The horizontal cells are coupled to each other through gap junctions, thus creating a resistive and capacitive sheet. This sheet therefore produces spatial and temporal averaging of the photodetector signals. Finally, the depolarizing and hyperpolarizing bipolar cells produce a differential signal that is fed to the retinal ganglion cells.

In addition, the photoreceptors and the neurons of the intermediate layers (the plexiform layers and the nuclear layer) produce signals with a very narrow dynamic range from widely varying amounts of incoming light. However, the output dynamic range is sufficient to capture an image's salient information.

This adaptation described above allows the retina to be compared to film in an analog camera. Adaptation is obtained through several processing techniques, such as a photoresponsive iris, pigment changes in the retinal receptors, and some neural processing in the visual cortex, and allows humans to appropriately respond to environmental scenes with illuminance that ranges over nine orders of magnitude, that is, from 10^{-4} lux (white objects under starlight, overcast sky, new moon) to 10^4 lux (the illuminance of a clear day). This feature of the visual system is most locally implemented by the OPL. The OPL allows a system with limited output range and finite analog resolution to communicate small local changes in image intensity when the background intensities may vary by a factor of one million.

From the intermediate processing layers, the transduced signal is finally represented using the brain's characteristic action potentials to ganglion cells, the axons of which form the optic nerve.

3.4.2
The Silicon Retina

Historically, the retina was among the first neural systems to be emulated in silicon, through the works of Mahowald and Mead [25, 33] and Boahen and Andreou [31]. As described in the previous section, three fundamental processing elements are required to construct a silicon retina:

- a phototransduction component;
- a network of horizontal cell-type elements, capable of spatially and temporally averaging the output of the phototransduction;
- a bipolar cell-type output, capable of amplifying the difference between the output of the phototransduction component and the horizontal cell signals.

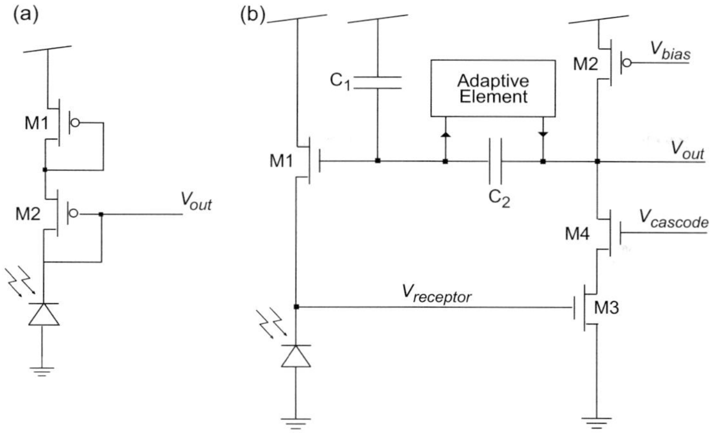

Figure 3.6 Components of a silicon retina. (a) Feedforward photoreceptor; (b) photoreceptor with adaptive feedback.

Physiological recordings show that the photoreceptors' electrical response is logarithmic in light intensity over the central part of the photoreceptors' range, as are the responses of other cells in the distal retina [30]. The logarithmic nature of the response has an important system-level consequence: the voltage difference between two points is proportional to their contrast ratio, that is, the difference between reflectance of two adjacent objects in an image.

To achieve this, as discussed in Equation 3.1, and with reference to Figure 3.6a, we can therefore connect a photodiode (note that in Mahowald's seminal work on the silicon retina [33], until shortly thereafter, p-type photo-bipolar junction transistors (BJTs) were used, that biase a series of diode-connected transistors in the subthreshold region to obtain an output voltage that is a function of the logarithm of the input current:

$$V_{\text{out}} \propto \ln\left(\frac{I_{\text{D}}}{I_0}\right) = \ln\left(\frac{I_{\text{photon}}}{I_0}\right). \tag{3.7}$$

This conversion to a log-domain space allows the compression of a large input range into a smaller output range. This simple circuit, known as a feed-forward retina [25], has no adaptation and its output voltage is typically too close to V_{DD} to be able to properly bias the component it is connected to, that is, the resistive network of horizontal cells. Hence, additional buffering or voltage-to-current conversion is required to connect to the resistive grid [34, 41].

The next evolution in the development of a neuromorphic photodetector was to add temporal adaptation to allow vision systems to operate over as wide a range of ambient intensity as their biological counterparts. The circuit in Figure 3.6b was developed by Delbruck and Mead [35] and has the following properties: (1) the load transistor (M_1) above the photodiode operates in subthreshold, implying that the voltage across the photo diode is logarithmic; (2) the voltage across the photodiode is amplified using a common-source cascaded, *inverting*, amplifier ($M_2 - M_4$); and (3) the output of the amplifier is band-pass filtered (by the adaptive element) and fed back to the load transistor M_1. Because it is fed back to the gate of M_1, which is a NMOSFET, it has the effect of readjusting the voltage across the photodiode to counter the output of the cascade amplifier. That is, the circuit realizes a negative feedback loop. Hence, if the photocurrent stays constant for some time (governed by the adaptive element), the output voltage (V_{out}) of the amplifier will return to close to its original value before the ambient light changed on the photodiode. This is adaptation, where the output amplifier magnifies changes to the photocurrent that are in the pass band of the adaptive element, but remain relatively insensitive to changes that are either too slow or too fast. As a result, the circuit was responsive to nearly *nine orders of magnitude* of ambient intensity, which is comparable to the human retina.

To achieve response times that were physiologically realistic (adaptation on the order of hundreds of milliseconds), the band-pass filter needed a particularly low frequency cutoff, which is hard to achieve in small VLSI circuits. Delbruck invented a device that used reverse biased MOS and parasitic BJT diodes, connected in parallel but opposite directions, to realize resistance as large as tens to hundreds of GΩ. Furthermore, these resistors were small, allowing them to be placed in each pixel of the retina. This was one of the key breakthroughs that made retinas with biologically comparable temporal adaptation possible.

3.4.3 Audition

The transduction of sound pressure signals into neural impulses forms the front end of the auditory system. Figure 3.7 shows a pictorial of the human ear, the organ responsible for detecting sound. Acoustic pressure waves enter the ear and are first reflected and attenuated by the pinna, then passed through the auditory canal at the end of which they impinge on the tympanic membrane. This acoustic pressure is then coupled with the cochlea's oval window (attached to the stapes) through the three middle-ear ossicles (malleus, incus, and stapes). The outer and middle ears, in other words, are responsible for gathering sound energy and efficiently coupling it into the inner ear.

The oval window, also shown in Figure 3.8, provides the acoustic input to the inner ear through the cochlea, a liquid-filled, nautilus shell-shaped organ.[1] When an acoustic pressure input, that is, a sound, enters the oval window, it displaces one fluid-filled cavity of the cochlea, known as the *scala media*. The moving fluid rests on a thin membrane, the *basilar membrane* (BM), which separates the scala media from an adjacent cavity, the *scala timpani*. The pressure wave that enters from the oval window will therefore propagate according to the characteristics of this membrane. If the membrane is stiff and thick, the wave propagates very rapidly, and is more sensitive to high frequency sound vibrations. If, on the other hand, the membrane is thin and flexible, then the pressure wave is slow, and resonates at lower frequencies. As it turns

1) The word "cochlea", in fact, comes from the Greek word for "shell."

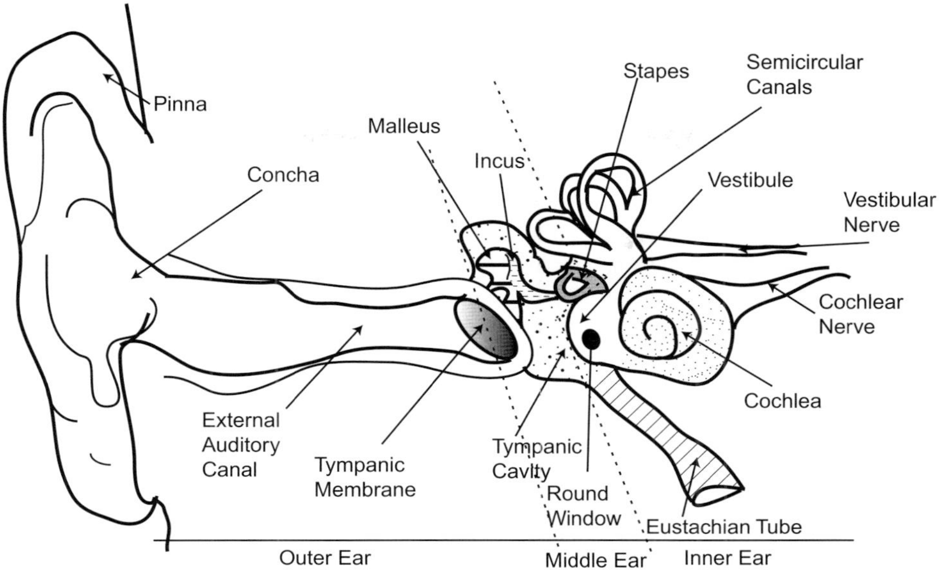

Figure 3.7 Pictorial of the human ear.

out, just like the strings on a piano, the basilar membrane is narrow and firm at its base (located at the oval window) and wide and compliant at its apex. To complete the circuit, the fluid in the scala media distorts the basilar membrane causing the fluid in the scala timpani to move toward the cochlea's exit, the round window.

The transduction itself from pressure wave to nerve signal is achieved by the hair cells, which detect the motion of the fluid on their cilia and responds with a change in the cell's voltage as well as release of neurotransmitter. The cochlea's basilar

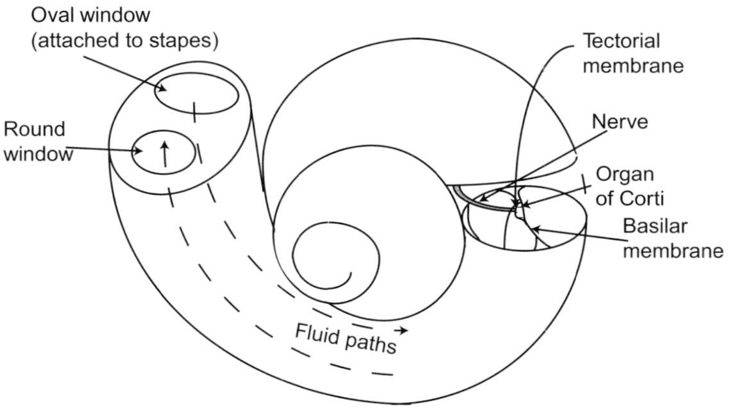

Figure 3.8 The cochlea is responsible for the transduction of the pressure signal representing the acoustic input into neural impulses that can be processed by the brain.

membrane is the hair cell's fixed base, such that each hair cell processes a differently filtered version of the pressure entering the ear. The fluid's displacement causes the row of 3500 hair cells (in humans) to move (1 Å is sufficient to be able to detect a sound), in turn causing nerve pulses to travel first to the cochlear nucleus and then to the brain.

Equally as important as the actual transduction is the ear's ability to adapt to a wide range of sound intensities through suppression of loud sounds and amplification of soft ones. The outer and middle ear attenuate loud sounds passively through directionality conformations and protective muscular reflexes to reduce transfer efficiency. The inner ear is responsible for amplification of soft sounds through the outer hair cells, which couple energy back into the system. When the sound signals are large, the outer hair cells are inhibited and the Automatic Gain Control (AGC) circuitry is effectively turned off. As the sounds become weaker, the inhibition is reduced and energy is fed back into the system.

3.4.3.1 Silicon Cochlea Modeling

Since all auditory processing starts with the cochlea, silicon auditory processing must seek to emulate the functionality of the cochlea. The cochlea's fundamental structure, as described in the previous section, is its basilar membrane. The characteristic frequency of the basilar membrane exponentially decreases from base to apex. This can be modeled by discretizing the BM into segments of equal length, where each segment is made up of a cascade of filters, roughly representing the inner hair cells, each with its own (exponentially decreasing) characteristic frequency. Specifically, in the original design for the silicon cochlea [36], the basilar membrane was modeled as a cascade of 480 stages of the second-order filters in a boustrophedonic arrangement to optimize the use of the silicon real estate available. The filters use wide-range OTAs, presented in Figure 3.9a. In Section 3.3, we derived the equation for an OTA, the transistors of which operated in the weak inversion (subthreshold) regime. Specifically, this was summarized with Equations 3.4 and 3.5, which showed that for

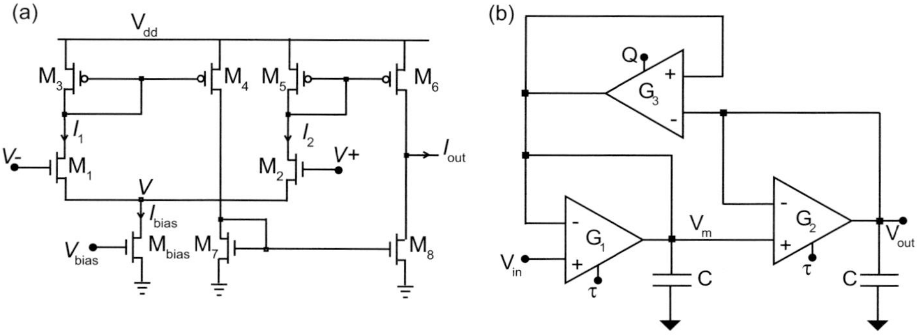

Figure 3.9 Second-order bandpass filters implemented to produce a 1D silicon cochlea. (a) Schematic of the wide-range OTA. In Lyon and Mead's implementation [36], 480 filters were used, each with a different (exponentially increasing) center frequency. (b) Sharp bandpass filter made with three wide-range OTAs.

small differences between the input voltages, the amplifier is linear with a transconductance $g_m = I_{bias}/U_T$, where I_{bias} is the amplifier's bias current and U_T is the thermal voltage. The wide-range OTA, depicted in Figure 3.9a, uses four more long transistors, M_5–M_8, to effectively maximize the output impedance of the device and thus its voltage gain.

Three wide-range OTAs are used to form a sharp (high-Q) bandpass filter: two of them (G_1 and G_2) are connected in a feed-forward, follower integrator configuration, while the third one (G_3) provides amplification through positive feedback (Figure 3.9b).

The follower integrators act as unity gain buffers at low frequencies, in which the output impedance of the circuit is very large, but act as low-pass filters/integrators at high frequencies. Specifically, at high frequencies, the integrator follower transfer function in the Fourier domain becomes

$$V_m = \frac{1/j\omega C}{1/G + 1/j\omega C} V_{in} = \frac{G}{j\omega C + G} V_{in} = \frac{1}{1 + j\omega C/G} V_{in}, \tag{3.8}$$

$$A_m = \frac{V_m}{V_{in}} = \frac{1}{1 + j\omega\tau}, \tag{3.9}$$

where G is the OTAs (tunable) transconductance, τ is the circuit's time constant, $\tau = C/G$, and A_m is the amplification of G_1 assuming G_2 and G_3 are off. This means that in the absence of feedback (G_3 is off), each integrator follower has a pole at $\omega = G/C$, after which the response falls at a rate of 6 dB/dec (i.e., 3 dB/dec per filter):

$$A_{out}\Big|_{G_3 = \text{off}} = \frac{V_{out}}{V_{in}} = \frac{1}{(1 + j\omega\tau)^2} \tag{3.10}$$

G_3 is an OTA in a positive feedback configuration that allows modulation of the quality factor, Q, of the second-order system. In general, Q characterizes the bandwidth of a resonator relative to its center frequency, $1/\tau$. Therefore, the small-signal transfer function for the filter with G_3 on becomes

$$A_{out} = \frac{1}{1 + 2j\omega\tau(1-\alpha) - \omega^2\tau^2} = \frac{1}{1 + (j\omega\tau/Q) - \omega^2\tau^2}, \tag{3.11}$$

where[2)]

$$\alpha = \frac{G_3}{G_1 + G_2} \tag{3.12}$$

and

$$Q = \frac{1}{2(1-\alpha)}. \tag{3.13}$$

The exponentially increasing time constant required to emulate the center frequency of the silicon basilar membrane was created by applying a voltage difference across a resistive polysilicon line [36]. This sets up a linear voltage gradient along the resistive line resulting in exponential changes in bias currents when using MOSFETs (in subthreshold regime).

2) Note that the above is also the equation of a second-order mechanical mass-spring system.

From the initial silicon cochlea, improvements took various different directions. The first modifications that were made to the circuitry aimed at improving matching in the second-order sections as well as the large-signal stability of the individual OTAs (e.g., Watts *et al.* [34] and van Schaik *et al.* [37]). Subsequently, Sarpeshkar *et al.* [38] further improved on the wide range OTAs by reducing the transconductance and also added a "fuse type" of nonlinearity that shuts off the positive-feedback amplification at large-signal levels instead of merely saturating it.

These approaches to designing the silicon cochlea are inherently imperfect: first, the serial cascade of second-order sections is such that if one of the sections fails, all the sections that follow become unusable; second, the number of second-order filter stages that can be employed is limited by the increasing delay between stages, which eventually prevents real-time usage of the system. These design issues spawned a different approach to the development of the silicon cochlea. Liu *et al.* [39, 40], for example, proposed a filter configuration in which the resonant elements (the band-pass filters) are placed in parallel to each other, whereas the cascaded elements (the low-pass filters) are either first-order or damped second-order filters, the delay of which could be varied independent of the characteristic frequencies of the resonators. It should be noted, however, that as with many issues associated with emulating biological systems, solutions to some potential issues with emulations come at the expense of the fidelity of the biological emulation itself.

3.5 Conclusions

This chapter provides an overview to help the reader understand the fundamental concepts of neuromorphic circuits. Starting with the brain's fundamental processing unit, the neuron, we showed how the underlying principles of the nervous system closely resemble those of MOS transistors biased in the subthreshold region, and how to emulate the basic functionality of the neuron, such as the production of the action potential, using simple circuitry that allows implementation of neuronal circuits on large scales.

The chapter also provided an overview of two important sensory systems and how basic CMOS circuits can rudimentarily emulate some of their characteristics. Specifically, we explained how biological systems confront the problem of transforming light into neural signals such that they can be processed by the brain, and how sound waves propagate through the ear canal and are converted to neural impulses by the fluid-filled cavities of the cochlea.

3.6 Resources

To implement neuromorphic circuits, tools for integrated circuits design are used. In particular, Cadence™ (a) and Tanner Tools™ (b) are the primary Electronic Design

Automation (EDA) packages used. They allow a user to perform schematic-level simulations of their circuits (using a physical circuit simulation software called SPICE), followed by drawing the physical layout of the circuit (i.e., define the material that must be placed on the silicon chip in order to realize the transistors and connections between them). Next, the layout and the drawn schematic must be compared to make sure that the layout is indeed an implementation of the schematic. This is done using a Layout versus Schematic (LVS) tool that is part of the packages. The final step is one where the schematic represented by the layout is Extracted (the name of the tool), with all parasitic capacitors, resistors, and diodes attached to the schematic. This extracted schematic can subsequently be simulated with SPICE to finalize the circuit functionality. This final simulation provides results that are usually the closest to the measured results after the chip is fabricated because it includes many of the parasitic devices that are usually not part of the initially drawn schematic.

Some neuromorphic designers have made some of their circuits available for download. Below we list two of these sites (c,d). Finally, there is an organization dedicated to the promotion of neuromorphic engineering and the dissemination of the community's products (e,f).

(a) Cadence Design Tools: www.cadence.com
(b) Tanner Design Tools: www.tannereda.com
(c) Tobi Delbruck's Neuromorphic Circuits Resources: www.ini.uzh.ch/~tobi/
(d) Kwabena Boahen's AER Circuits Resources: www.stanford.edu/group/brainsinsilicon/boahen.html
(e) The Institute for Neuromorphic Engineering: www.ine-web.org
(f) Neuromorphic Engineering and Cognition Blog: www.neurdon.com

References

1 Thrun, S., Montemerlo, M., Dahlkamp, H., Stavens, D., Aron, A., Diebel, J., Fong, P., Gale, J., Halpenny, M., and Hoffmann, G. (2006) Stanley: the robot that won the DARPA grand challenge. *J. Field Robotics*, **23**, 661–692.

2 Boahen, K.A. (2000) Point-to-point connectivity between neuromorphic chips using address events. *IEEE Trans. Circuits II*, **47** (5), 416–434.

3 Cassidy, A. and Andreou, A. (2008) Dynamical digital silicon neurons. Proc. of IEEE Biomedical Circuits and Systems Conference, Baltimore, November.

4 Mead, C. (1989) *Analog VLSI and Neural Systems*, Addison-Wesley, Reading, MA.

5 Grey, P. and Meyer, R. (2001) *Analysis and Design of Analog Integrated Circuits*, 4th edn, John Wiley & Sons, Inc., MA.

6 Sedra, A. and Smith, K. (2004) *Microelectronics*, 5th edn, Oxford University Press, New York.

7 Vittoz, E. (2002) Present and future industrial applications of bio-inspired VLSI systems. *Analog Integr. Circ. Signal Porcess.*, **30** (2), 173–184.

8 Mahowald, M. (1992) *VLSI Analogs of Neuronal Visual Processing: A Synthesis of Form and Function*, California Institute of Technology.

9 Morris, C. and Lecar, H. (1981) Voltage oscillations in the barnacle giant muscle fiber. *Biophys. J.*, **35**, 193–213.

10 Smith, L.S. (2006) Implementing neural models in silicon, in *Handbook of Nature-Inspired and Innovative Computing: Integrating Classical Models with Emerging Technologies* (ed. A. Zomaya), Springer, pp. 433–475.

11 Izhikevich, E. (2004) Which model to use for cortical spiking neurons? *IEEE Trans. Neural Netw.*, **15**, 1063–1070.
12 Mahowald, M. and Douglas, R. (1991) A silicon neuron. *Nature*, **354**, 515–518.
13 Patel, G.N. and DeWeerth, S.P. (1997) Analog VLSI Morris-Lecar neuron. *Electro. Lett.*, **33** (12), 997–998.
14 Simoni, M.F., Cymbalyuk, G.S., Sorensen, M.E., Calabrese, R.L., and DeWeerth, S.P. (2004) A multi-conductance silicon neuron with biologically matched dynamics. *IEEE Trans. Biomed. Eng.*, **51** (2), 342–354.
15 van Rossum, M.C.W., Bi, G.Q., and Turrigiano, G.G. (2000) Stable Hebbian learning from spike timing-dependent plasticity. *J. Neurosci.*, **20** (23), 8812–8821.
16 Arthur, J. and Boahen, K. (2006) Learning in silicon: timing is everything in *Advances in Neural Information Processing Systems 18* (eds Y. Weiss, B. Scholkopf, and J. Platt), MIT Press, Cambridge, MA, pp. 75–82.
17 Arthur, J.V. and Boahen, K. (2006) Silicon neurons that inhibit to synchronize. Proceedings of the IEEE International Symposium on Circuits and Systems.
18 Rasche, C. and Douglas, R.J. (2001) Forward- and backpropagation in a silicon dendrite. *IEEE Trans. Neural Netw.*, **12** (2), 386–393.
19 Hasler, P. and Lande, T.S. (2001) Overview of floating-gate devices, circuits, and systems. *IEEE Trans. Circuits II*, **48** (1), 1–3.
20 Berlucchi, G. (1999) Some aspects of the history of the law of dynamic polarization of the neuron: from William James to Sherrington, from Cajal and van Gehuchten to Golgi. *J. Hist. of Neurosci.*, 8 (2), 191–201.
21 Indiveri, G. (2003) A low-power adaptive integrate-and-fire neuron circuit. Proceedings of the IEEE International Symposium on Circuits and Systems, pp. IV820–IV823.
22 Indiveri, G., Chicca, E., and Douglas, R. (2006) A VLSI array of low-power spiking neurons and bistable synapses with spike-timing dependent plasticity. *IEEE Trans. Neural Netw.*, **17** (1), 211–221.
23 Folowosele, F., Harrison, A., Cassidy, A., Andreou, A.G., Etienne-Cummings, R., Mihalas, S., Niebur, E., and Hamilton, T.J. (2009) A switched-capacitor implementation of the generalized linear integrate-and-fire neuron. Proceedings of the IEEE International Symposium on Circuits and Systems.
24 Mihalas, S. and Niebur, E. (2009) A generalized linear integrate-and-fire neural model produces diverse spiking behaviors. *Neural Comput.*, **21** (3), 704–718.
25 Mahowald, M. (1992) VLSI Analogs of Neuronal Visual Processing: A Synthesis of Form and Function. PhD thesis, California Institute of Technology.
26 Goldberg, D.H., Cauwenberghs, G., and Andreou, A.G. (2001) Probabilistic synaptic weighting in a reconfigurable network of VLSI integrate-and-fire neurons. *Neural Netw.*, **14** (6–7), 781–793.
27 Vogelstein, R.J., Mallik, U., Culurciello, E., Cauwenberghs, G., and Etienne-Cummings, R. (2007) A multichip neuromorphic system for spike-based visual information processing. *Neural Comput.*, **19**, 1–20.
28 Choi, T.Y.W., Shi, B.E., and Boahen, K.A. (2005) Neuromorphic implementation of orientation hypercolumns. *IEEE Trans. Circuits I*, **52** (6), 1049–1060.
29 Culurciello, E., Etienne-Cummings, R., and Boahen, KA. (2003) A biomorphic digital image sensor. *IEEE J. Solid State Circuits*, **38** (2), 281–294.
30 Dowling, J.E. (1987) *The Retina: An Approachable Part of the Brain*, Harvard University Press.
31 Boahen, K. and Andreou, A.G. (1991) A contrast sensitive silicon retina with reciprocal synapses. *Adv. Neural Inform. Process. Syst.*, Morgan Kaufmann, pp. 764–772.
32 Mahowald, M. (1994) *An Analog VLSI System for Stereoscopic Vision*, Kluwer Academic Publishers, Boston, MA.
33 Mahowald, M. and Mead, C. (1991) The silicon retina. *Scientific American*, **264**, 76–82.
34 Watts, L., Kerns, D.A., Lyon, R.F., and Mead, C.A. (1992) Improved implementation of the silicon cochlea. *IEEE J. Solid State Circuits*, **25** (5), 692–700.
35 Delbruck, T. (2003). Investigations of Analog VLSI Visual Phototransduction and Motion Processing, Ph.D. thesis, California Institute of Technology, Pasadena, CA.

36 Lyon, R.F. and Mead, C. (1988) An analog electronic cochlea. *IEEE Trans. Acoust.*, **36** (7), 1119–1134.

37 van Schaik, A., Fragniere, E., and Vittoz, E. (2005) Improved silicon cochlea using compatible lateral bipolar transistors, in *Advances in Neural Information Processing Systems 8*, MIT Press, Cambridge, MA.

38 Sarpeshkar, R., Lyon, R.F., and Mead, C. (1998) A low-power wide-dynamic-range analog VLSI cochlea. *Analog Integr. Circ. Signal Process.*, **16** (3), 245–274.

39 Liu, W., Andreou, A.G., and Goldstein, M.H. (1991) An analog integrated speech front-end based on the auditory periphery. Proceedings of International Joint Conference on Neural Networks, pp. 861–864.

40 Liu, W., Andreou, A.G., and Goldstein, M.H. (1992) Voiced-speech representation by an analog silicon model of the auditory periphery. *IEEE Trans. Neural Netw.*, **3** (3), 477–487.

41 Etienne-Cummings, R. and Van der Spiegel, J. (1996) Neuromorphic vision sensors. *Sens. Actuators A*, **56** (1/2), 19–29.

4
Principles of Neural Signal Processing

Don H. Johnson

4.1
Introduction

Sensory neurons represent information about the environment and motor neurons represent commands to muscular systems. Clearly, the neural codes employed by various systems operate under different constraints; as a consequence, the representation of information within each system varies considerably. For example, neurons in the visual and auditory systems face different demands and encode information differently. As a result, *the* neural code does not exist. Neural signal processing endeavors to determine how neurons represent information and with what accuracy. Biohybrid systems go even further, attempting either to create neural signals to restore sensorimotor function (what we call stimulation prosthetics) or to extract information from neural signals to generate control signals (measurement prosthetics).

Because of the stochastic nature of most neural spike trains, the mathematical tool used to describe spike trains as a spatiotemporal sequence of isolated events is *point process theory* [1–3]. After using this theory to describe what neural coding is, we describe the signal processing techniques neuroscientists use to determine the neural code and engineers use to extract the coded information from recorded signals. The discussion then turns to an information theoretic description of the limits to which neurons can represent sensory or motor signals. This approach also provides insights into the consequences of simple biohybrid interfaces for prosthetic applications.

4.2
Point Process Theory

4.2.1
Definition of a Point Process

A *regular point process* [1] is defined so that the probability of an event occurring in the time interval $[t, t+\Delta t)$ is given by

Biohybrid Systems: Nerves, Interfaces, and Machines, First Edition. Edited by Ranu Jung.

$$\begin{aligned} \Pr[\text{one event in } [t, t+\Delta t) | N_t, \mathbf{w}_t] &= \mu(t; N_t, \mathbf{w}_t)\Delta t, \\ \Pr[\text{more than one event in } [t, t+\Delta t) | N_t, \mathbf{w}_t] &= o(t, \Delta t). \end{aligned} \tag{4.1}$$

Here, N_t is the number of events that have occurred prior to time t (observations are assumed to start at time $t = 0$); $\mathbf{w}_t$ is the vector of occurrence times of these N_t events: $\mathbf{w}_t = [w_1, \ldots, w_{N_t}]$; and $o(t, \Delta t)$ decreases to zero with decreasing Δt faster than linearly: $\lim_{\Delta t \to 0} o(t, \Delta t)/\Delta t = 0$. These equations mean that no more than one event can occur in a sufficiently small interval and that the probability of one event occurring within a small interval is proportional to the interval duration. Most important to this definition is the *intensity* $\mu(t; N_t, \mathbf{w}_t)$ that governs how the event probability depends on time and on the process's *history* as encapsulated by N_t and $\mathbf{w}_t$: the number and times at which all events occurred prior to time t. These quantities are depicted in Figure 4.1. Note that the probabilities are *conditional* probabilities: they all depend on the point process's history. The conditional probability of an event occurring during a small time interval equals the product of the interval duration Δt and the nonnegative $\mu(t; N_t, \mathbf{w}_t)$. The intensity has units of events per second and can

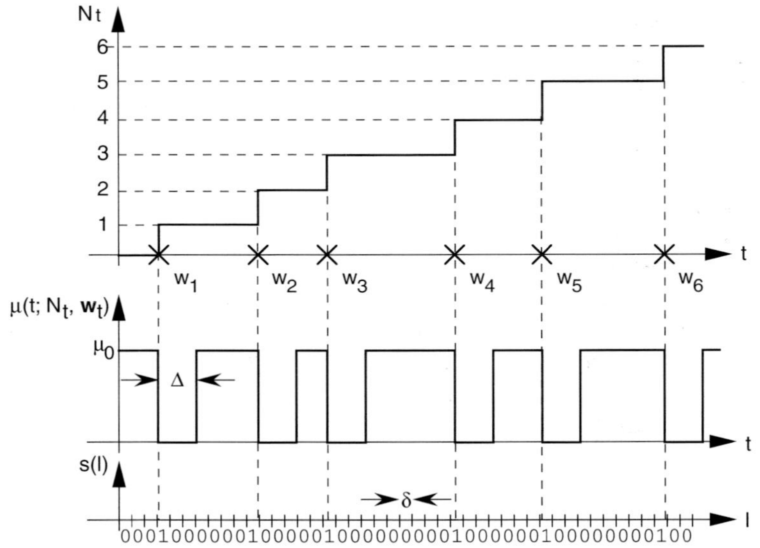

Figure 4.1 The essential quantities that define a point process are noted along side a sample function in the upper panel. w_n is the occurrence time of the nth event and $\tau_n = w_{n+1} - w_n$ is the interevent interval. N_t is known as the counting process. The middle trace depicts the corresponding intensity of a refractory process. This intensity depends on immediate process history, with the intensity dropping to zero for a time interval equaling Δ after each event. This type of intensity means that event occurrence is suppressed for a period of time after each event. The bottom trace shows the usual procedure for the analog-to-digital conversion of a spike train: divide time into intervals of duration δ and produce the number of events that occur within each interval.

be considered the *instantaneous rate* at which events occur. The intensity can depend on time and/or the point process's history. If the intensity depends explicitly on time t, the process is nonstationary. Such temporal variations usually represent rate variations that are due to external influences and have been suitably modified – processed – by the neuron. Such variations represent how information is encoded in the point process. The "historical" portion of the intensity describes the detailed interactions of previous events on the current one. In a neuroscientific context, this history could, among other possibilities, represent the neuron's dynamic active-membrane properties and its sensitivity to modulators and transient ionic concentrations, all of which certainly contribute to the neuron's processing function. For example, the discharge rate typically decreases to zero momentarily after an action potential occurs. This refractory effect has long been noted; in the present context, this phenomenon means that the intensity for virtually all accurate point process models of neural discharges must contain a historical component. Consequently, both the neuron's biophysics and the encoded information affect the discharge behavior of a neuron.

Once we have the intensity of a point process, the process's entire statistical structure is specified. Consequently, a thorough analysis of a discharge pattern must yield the process's intensity function. In technical terms, deciphering the neural code amounts to measuring the intensity and teasing apart the intrinsic and extrinsic components. Even when the rate is constant, a seemingly boring situation from an information processing viewpoint, careful analysis of the intensity's historical component can yield important insights into the neuron's processing function when rate variations are present. Unfortunately, the intensity is usually related to neural mechanisms in a complicated, not easily deciphered, way.

4.2.2 Examples of Point Processes

4.2.2.1 The Poisson Process

The Poisson process has the simplest structure of any point process. Its intensity does not depend on history: $\mu(t; N_t, \mathbf{w}_t) = \lambda(t)$. The quantity $\lambda(t)$ could be constant, meaning that events occur randomly at a fixed average rate (the process is *stationary*), it could vary with time, meaning that occurrence rate varies accordingly, or it could vary stochastically, resulting in what is known as a *doubly stochastic Poisson process*. No matter which of these situations applies, event occurrence does not depend statistically on when previous events occur: the Poisson process has no memory. Since a neuron's dynamic membrane properties would most certainly introduce memory into its output, the Poisson process can approximate only neural output.

In point process theory, the Poisson process is akin to white Gaussian noise in random process theory: analytic results about the Poisson process's structure are easy to derive and virtually all other point processes present analytic difficulties. For example, the probability that n events occur between the times t and $t + T$, $T > 0$, is given by [1, 3]

$$\Pr[N_{t+T}-N_t = n] = \frac{1}{n!}\left(\int_t^{t+T} \lambda(\alpha)\,\mathrm{d}\alpha\right)^n \exp\left\{-\int_t^{t+T} \lambda(\alpha)\,\mathrm{d}\alpha\right\}, n = 0, \ldots \tag{4.2}$$

We define the *rate of discharge* $\hat{\lambda}(t)$ to be the limit of the expected number of events in an interval that begins at time t divided by the interval's duration:

$$\hat{\lambda}(t) = \lim_{\Delta t \to 0+} \frac{E[N_{t+\Delta t}-N_t]}{\Delta t}.$$

Here, $E[\cdot]$ denotes expected value. For the Poisson point process, the expected value equals $\int_t^{t+\Delta t} \lambda(\alpha)\,\mathrm{d}\alpha$, and the discharge rate equals the smoothed intensity as expected. We shall find that the Poisson point process is *unique* in having this property. We can derive from (4.2) the probability density function (pdf) of the interevent interval τ_n.

$$p_{\tau_{N_t+1}}(\tau; t) = \lambda(w_{N_t}+\tau)\exp\left\{-\int_{w_{N_t}}^{w_{N_t}+\tau} \lambda(\alpha)\,\mathrm{d}\alpha\right\}. \tag{4.3}$$

When the Poisson process is stationary, with $\lambda(t) = \lambda_0$, we have the well-known result that the interevent intervals in a Poisson process are exponentially distributed: $p_\tau(\tau) = \lambda_0 \exp\{-\lambda_0 \tau\}, \quad \tau > 0$.

Despite its limitations to describe single-neuron data, the Poisson process provides a starting point for understanding a given point process's properties. The rule of thumb is that if you cannot analyze a Poisson process approximation, related analysis of a more exact point process description will almost certainly be more difficult, if not impossible.

4.2.2.2 Renewal Processes

A *renewal* process results when the intensity's historical dependence amounts to the time at which the last event preceding time t occurred [4]: $\mu(t; N_t, \mathbf{w}_t) = \mu(t; N_t, w_{N_t})$. Thus, interevent *intervals* are statistically independent random variables, with the probability of an event occurring varying with time since the last event. When the intensity does not depend on absolute time, the intensity is a function only of the interval $t-w_{N_t}$ since the last event:

$$\mu(t; N_t, \mathbf{w}_t) = H(t-w_{N_t}). \tag{4.4}$$

If the function $H(\cdot)$ is a constant (no dependence on the interval since the last event), a Poisson process results. If this function equals zero for small arguments and then jumps to some nonzero value, $H(\tau) = \mu_0 \mathrm{u}(\tau-\Delta)$, we can describe absolute refractory effects with Δ identified as the refractory interval and $\mathrm{u}(\cdot)$ is the unit step function. After the refractory interval has ended, the constant-valued intensity means that the process has entered a Poisson-like mode, with event occurrence oblivious to event history. The lower panel of Figure 4.1 depicts the intensity for this type of renewal process.

This figure also clarifies the somewhat confusing point that the average occurrence rate for this so-called *deadtime-modified* point process does *not* equal μ_0; the rate of event occurrence must be less than μ_0. To compute the average rate $\bar{\lambda}$, note that it should equal the long-term average of the intensity. Considering Figure 4.1, the area under the intensity equals $\mu_0 T(1-\bar{\lambda}\Delta)$. Dividing this equation by T and setting the result equal to $\bar{\lambda}$ yield $\bar{\lambda} = \mu_0/(1+\mu_0\Delta)$. This example demonstrates that the intensity function indicates the *true* rate of discharge immediately following an event while the average rate indicates the overall rate at which events occur. Another conclusion can be drawn: gross measures of event rate – ones that do not take event history into account – do not indicate the true rate within which information is encoded. However, we can, in some circumstances, determine underlying quantities from measured ones. Assuming we know the value of Δ, we could estimate a renewal process's true rate variations [5, 6].

Every stationary renewal process is completely characterized by the probability density function $p_\tau(\cdot)$ of the interval between successive events. This fact follows because the procedure used to derive (4.3) can be used on renewal processes. This pdf is related to the process's intensity through the function $H(\cdot)$ defined in (4.4) as

$$\begin{aligned} H(\tau) &= \frac{p_\tau(\tau)}{\int_\tau^\infty p_\tau(\alpha)\,d\alpha}, \\ p_\tau(\tau) &= H(\tau)\exp\left\{-\int_0^\tau H(\alpha)\,d\alpha\right\}. \end{aligned} \tag{4.5}$$

The quantity $H(\cdot)$ is known as the *hazard function* or the *recovery function* associated with the point process. Clearly, this quantity is just the renewal process's intensity. Our example has $H(\tau) = \mu_0 u(\tau-\Delta)$, which corresponds to the interval pdf $p_\tau(\tau) = \mu_0 \exp\{-\mu_0 \cdot (\tau-\Delta)\}u(\tau-\Delta)$.

Calculating the hazard function frequently provides more insight into the process's structure than does the interval distribution. The hazard function $H(\tau)$ is proportional to the probability of an interval equaling τ given that the interval is at least that long; this interpretation follows directly from (4.5). A constant-valued hazard function means that the probability of an event occurring is as likely now as it is later: the process has no memory over the range of intervals for which the hazard function, the intensity, is constant. If the hazard function is an increasing function, then the process becomes more "anxious" to produce an event as the interval becomes longer. For example, if the hazard function increases linearly, $H(\tau) = K\cdot\tau$, K a constant, then $p_\tau(\tau) = K\tau\exp\{-K\tau^2/2\}$, a much narrower distribution of intervals than in the Poisson (a constant hazard function $\leftrightarrow$ exponential interval pdf) case. A decreasing hazard function has the opposite interpretation. However, a point process's intensity cannot decrease to zero as τ approaches infinity; if it does, events could cease to occur, and the process would not be self-consistent.[1)]

1) More precisely, a valid hazard function's integral must be infinite: $\int_0^\infty H(\alpha)\,d\alpha = \infty$. In most cases, $H(\cdot)$ is bounded; however, this technical condition allows unbounded choices.

4.2.2.3 Markov Point Processes

Events in nonrenewal processes depend not only on when the previous event occurred but also on ones before that. In *Markov* point processes, event occurrence depends on the most recent $k+1$ events and not on ones prior to those. Consequently, the intensity simplifies to some degree.

$$\mu(t;\,N_t,\mathbf{w}_t)=\mu(t;\,N_t,[w_{N_t},w_{N_t-1},\ldots,w_{N_t-k}]).$$

The number k is known as the *order* of the Markovian dependence structure. A renewal process has order zero: event occurrence depends only on the most recent event. Higher-order dependence means that successive interevent intervals are correlated random variables, which complicates development of concise descriptions of data.

If the spike production mechanism is described "classically" – a synaptic event influences the transmembrane potential of active membrane modeled by Hodgkin–Huxley Na^+ and K^+ channels – the equations governing spike production have no state variable that persists beyond the occurrence of an action potential. Thus, spike occurrence does depend on when the previous spike occurs but not on any others, which means that a renewal process model *always* results.

4.2.2.4 Non-Markovian Point Processes

In many neuron types, some persistent variable (such as Ca^{2+}) affects active membrane's dynamics and the consequent production of action potentials. The dynamic equations that describe these phenomena indicate that spike production depends not on some number of previous events (which would lead to a Markov model) but on a time-weighted average. For example, the intensity at time t would depend on the output of a linear filter having an input equal to impulses occurring at all previous event times [7]:

$$\mu(t;\,N_t,\mathbf{w}_t)=\lambda_0+\int_0^t g(t-\alpha)\,\mathrm{d}N_\alpha.$$

The Stieltjes representation of this convolution integral is needed for technical reasons; it simply equals the summed impulse responses delayed by all event times occurring prior to time t:

$$\mu(t;\,N_t,\mathbf{w}_t)=\lambda_0+\sum_{i=1}^{N_t} g(t-w_i).$$

Thus, the intensity depends on *all* past event times viewed through the window described by $g(\cdot)$. This model could be used to describe situations in which event occurrence depends on the times at which all previous events occur (when $g(\cdot)$ has infinite duration) or that it varies with events occurring within the most recent T seconds ($g(\cdot)$ is nonzero only over $[0,T]$).

An arbitrary impulse response $g(\cdot)$ cannot be used in this expression. For instance, intensities are always positive quantities, meaning that the summed impulse responses cannot be more negative than $-\lambda_0$. Further restrictions on the impulse response result if we demand this so-called Hawkes' process be stationary. Defining $\bar{\lambda}$ as the average occurrence rate, the intensity must satisfy

$$\bar{\lambda} = \lambda_0 + \bar{\lambda} \int_{-\infty}^{t} g(t-\alpha)\, d\alpha \quad \Rightarrow \quad \bar{\lambda} = \frac{\lambda_0}{(1 - \int_0^\infty g(\alpha)\, d\alpha)}.$$

This result means that the filter's impulse response must satisfy $\int_0^\infty g(\alpha)\, d\alpha < 1$ for the Hawkes process to exist. The Hawkes' process is frequently used in the so-called *log-linear* models of point process intensity, which allows much simpler use of maximum likelihood and Bayesian techniques for analyzing point processes [8, 9].

4.2.3
Multiple Point Processes

Experimental techniques today allow several spike trains to be recorded simultaneously, either with individual electrodes that are spatially separated or with electrode arrays [10]. Rather than recording a single neuron's pattern, each electrode provides a superposition of spike trains from an unknown number of neurons. As we will show later, data provided this way is of little use for prosthetic applications. The so-called *spike sorting* algorithms endeavor to recast such recordings into individual neuron spike trains. In the single electrode case, optimal classification algorithms exploit spike amplitude and waveform differences to separate a single recording into multiple point processes representing individual neuron spikes [11]. When electrode arrays are used, several electrodes can record the same neuron's signal. Array processing algorithms that exploit event simultaneity, as well as amplitude and waveform, separate multiple recordings into a multichannel point process [12]. The article by Lewicki [13] surveys spike sorting issues and algorithms.

Describing several point processes as jointly defined random processes is quite complicated. The joint distribution of more than two Poisson processes has only been conjectured [14, 15]. What has been shown is that jointly defined Poisson processes can be dependent only positively: the occurrence of an event in one process increases the probability of an event in another. Both positively and negatively dependent simultaneously recorded spike trains have been observed. Negatively correlated pairs of renewal processes can be defined [16], but not in a way extensible to more. Furthermore, triple and higher order dependencies only loosely related to pairwise dependencies must be present among jointly defined Poisson processes [15].

One way of defining the joint distribution of several point processes is as *marked* point processes [1]. Here, each event carries a mark – a value – that defines which process produced it. While such a description is possible, it has not yet proven its utility.

4.3
Analyzing a Point Process

Point processes are no different from other signals in at least one way: before computer analysis is possible, the process (or processes) must be digitized. Here, the "analog-to-digital" conversion, illustrated in Figure 4.1, consists of counting the number of events that occur in a sequence of small intervals.

$$s(l) = N_{l\delta_s,(l+1)\delta_s} \triangleq N_{(l+1)\delta_s} - N_{l\delta_s}.$$

In most cases, the *binwidth* δ_s is chosen to be less than the refractory interval, which means the digitized spike train, denoted by $s(l)$, is binary valued. Consequently, the point process is converted into a Bernoulli process wherein the probability of an event occurring equals

$$\Pr[s(l) = 1] = \int_{l\delta_s}^{(l+1)\delta_s} \mu(t; N_t; \mathbf{w}_t)\, \mathrm{d}t,$$

thereby mimicking (4.1) when δ_s is chosen to be smaller than the temporal variations of the intensity.

When multiple spike trains are digitized, the bins are aligned and a common binwidth is usually chosen. Each "sample" now consists of a vector having a length equal to the number of neurons being digitized M and with each component equaling the number of events occurring during the bin from one of the spike trains. Consequently, we now have a vector-valued process $\mathbf{s}(l)$ in which both temporal and spatial (along the dimension of each vector) variations and/or statistical dependencies can occur. Only when binary-valued elements in each vector sample occur do we have a viable probabilistic description, a vector Bernoulli process [14]. This model can describe statistical dependencies in the spatial dimension but not in time.

The sampling process has a signal processing model [17]. First of all, the model represents each point process by a sequence of delta functions according to $\mathrm{d}N_t/\mathrm{d}t$. The "analog-to-digital" conversion process for point processes is equivalent to passing this sequence through a finite-time integrator (duration δ_s), then sampling every δ_s. The impulse response $h_{\delta_s}(t)$ of the finite-time integrator is

$$h_{\delta_s}(t) = \begin{pmatrix} 1 & 0 < t < \delta_s, \\ 0 & \text{otherwise.} \end{pmatrix}$$

The transfer function of this system is proportional to sinc $(\pi f \delta_s)$, which has a low-pass character. Thus, counting the number of events that occur in a bin amounts to a nonideal antialiasing filter having a cutoff frequency somewhat less than $1/2\delta_s$. The subsequent sampling, $s(l) = (\mathrm{d}N_t/\mathrm{d}t) \otimes h_{\delta_s}(t)|_{t=l\delta_s}$, that produces the digitized point process can introduce aliasing if the intensity varies more rapidly than the reciprocal binwidth.

Digitizing several jointly recorded spike trains is no different from applying the binning procedure to each spike train individually. As we will learn later, this process imposes additional constraints beyond those predicted by sampling theory.

4.3.1 The Interval Histogram and Hazard Function

The interval histogram estimates the probability density function of interevent interval durations. When the process is stationary and renewal, the interval histogram suffices to characterize the process. When interval dependencies are present or when the process is nonstationary, the interval histogram no longer suffices to describe the process. Nevertheless, the interval histogram and the associated hazard function are frequently measured to assess the spike train's intrinsic statistical properties.

Algorithmically, the interval histogram computes the fraction of measured interevent intervals having durations between $l\delta$ and $(l+1)\delta$, then normalizes by the binwidth so that the estimate "integrates" to one. Note that the interval histogram's binwidth δ is usually larger than the sampling binwidth δ_s. Letting $I_{l,\delta}(\cdot)$ denote the indicator function,

$$I_{l,\delta}(\tau) \equiv \begin{cases} 1, & l\delta \leq \tau < (l+1)\delta, \\ 0, & \text{otherwise}, \end{cases}$$

the interval histogram computation is expressed by [17]

$$\mathrm{INT}(l) = \frac{1}{N\delta} \sum_{n=0}^{N-1} I_{l,\delta}(\tau_n), \tag{4.6}$$

where N is the total number of intervals entering into the computation. The expected value of the histogram equals $\int_{l\delta}^{(l+1)\delta} p_\tau(\tau)\,\mathrm{d}\tau/\delta$. If the true interval distribution is essentially constant over all length-δ bins, this expected value equals the interval pdf evaluated at $l\delta$: $\mathrm{E}[\mathrm{INT}(l)] \approx p_\tau(l\delta)$. Assuming a renewal point process, we can easily analyze the variance of the interval histogram. The variance of a sum of N independent binary-valued random variables equals $Np(1-p)$. Assuming that $p \ll 1$, the variance of the interval histogram for a renewal process equals

$$\mathrm{V}[\mathrm{INT}(l)] = \frac{\int_{l\delta}^{(l+1)\delta} p_\tau(\tau)\,\mathrm{d}\tau}{N\delta^2} \approx \frac{p_\tau(l\delta)}{N\delta}.$$

To assess this variance better, we consider the histogram's coefficient of variation, which measures the percentage variation of nonnegative-valued random variables as the ratio of the random variable's standard deviation and its expected value.

$$\mathcal{C}[X] \triangleq \frac{(V[X])^{1/2}}{\mathrm{E}[X]}. \tag{4.7}$$

Thus, the coefficient of variation for each bin of the interval histogram is

$$\mathcal{C}[\mathrm{INT}(l)] = 1\Big/\sqrt{N\int_{l\delta}^{(l+1)\delta} p_\tau(\tau)\,\mathrm{d}\tau} \approx [N\delta p_\tau(l\delta)]^{-1/2}.$$

Thus, the percentage error increases as the density's value decreases: we can more accurately estimate the histogram's modes than we can its valleys and tail.

Because the hazard function and the interval pdf are directly related, we estimate the hazard function using a discrete version of (4.5):

$$\mathrm{HAZ}\,(l) = \frac{\mathrm{INT}(l)}{\sum_{k=l}^{\infty} \mathrm{INT}(k)\delta}.$$

Calculating the coefficient of variation for the hazard function is analytically difficult because the hazard function is the ratio of two random, correlated quantities. Using an approximation detailed in Johnson [3], we find that

$$\mathcal{C}[\mathrm{HAZ}(l)] \approx \sqrt{\frac{1}{N\delta p_\tau(l\delta)} - \frac{1}{N\delta \sum_{k=l}^{\infty} p_\tau(k\delta)}} = \mathcal{C}[\mathrm{INT}(l)][1-H(l\delta)\delta]^{1/2}.$$

This expression is meaningless unless $H(l\delta)\delta \ll 1$, which means that the binwidth is much less than the smallest interspike interval. In such situations, the coefficient of variation for the hazard function estimate approximately equals that for the interval histogram, with both becoming large for the same intervals. However, as opposed to the interval histogram, the value of the hazard function is usually not small when these large values occur, which accentuates the errors.

4.3.2
The PST Histogram

The PST (poststimulus or peristimulus time) histogram measures the temporal variations in a single neuron's rate of discharge by averaging the spike-train response to a periodically repeating stimulus. The onset of each stimulus presentation serves as the histogram's time origin. Assume that the time axis is quantized into bins of length δ (an integer multiple of the sampling binwidth) and that the stimulus is presented every R bins with the first stimulus presented at time $t = 0$. The count of the events in the rth bin $[r\delta, (r+1)\delta)$ is denoted by $c(r)$. If the analysis binwidth equals the sampling binwidth, $\delta = \delta_s$, the event count equals the sampled spike train. The PST histogram computation averages this sequence across stimulus presentations [17]:

$$\mathrm{PST}(r) = \frac{1}{K\delta}\sum_{k=0}^{K-1} c(r+kR).$$

Here, K denotes the number of stimulus presentations. Normalized in this way, the units of a PST histogram are events/s.

For even the simplest point process, the expected value of the PST histogram contains many surprises. First, assume that the binwidth is small so that the intensity definition of (4.1) can be used. This assumption means that more than one event rarely occurs within a single bin. Under this condition, the expected value of the sequence $c(r)$ is $\mathrm{E}[\mu(r\delta; N_{r\delta}, \mathbf{w}_{r\delta})]\delta$, where the expected value is computed with respect to event history. When the intensity's time dependence repeats with period $R\delta$, the expected value of a PST histogram equals this quantity divided by δ. Thus, the PST histogram corresponds to averaging over all event histories that lead to or prevent an event occurring at time $r\delta$ after the stimulus presentation:

$$\mathrm{E}[\mathrm{PST}(r)] = \mathrm{E}[\mu(r\delta; N_{m\delta}, \mathbf{w}_{r\delta})]. \tag{4.8}$$

Thus, this expected value does *not* depend solely on the intensity's temporal variation. The *only* point process that expresses this temporal dependence directly is the Poisson because its intensity does not depend on event history. For all other point processes, including renewal ones, *the PST histogram represents stimulus-related temporal variations averaged across the point process's history*. This expected value is extremely difficult to evaluate. Consequently, the PST histogram is a complicated function of the process's stimulus-related nonstationarity and the inherent "nonstationarity" induced by the process's statistical structure [5].

4.3.3 Characterizing Multiple Point Processes

The interval and response characteristics of the individual point processes of an ensemble recording can be analyzed as described previously. But using only these marginal measures to characterize multiple point processes implicitly assumes that the processes are statistically independent. Such a strong assumption needs verification. In general, assessing whether two random variables, much less a larger number, are statistically independent is a difficult problem. What makes this assessment even more difficult in neural recordings are nonstationarities, which are always likely to be present. When these time variations can be linked to stimulus presentations, the statistical accuracy of any dependence test depends on the number of stimulus repetitions. Statistical demands require this number to be related *exponentially* to the number of point processes being considered, overwhelming the capabilities of most experiments.

Though usually a weak measure of statistical dependence, the correlation coefficient function $\varrho_{ij}(\ell)$ between pairs of digitized processes $s_i(\cdot)$, $s_j(\cdot)$ expresses whether the digitized processes are statistically independent or not. Extending the usual definition of correlation coefficient to include a lag ℓ,

$$\varrho_{ij}(l, \ell) = \frac{\mathrm{E}[(s_i(l+\ell) - \mathrm{E}[s_i(l+\ell)]) \cdot (s_j(l) - \mathrm{E}[s_j(l)])]}{[\mathrm{V}\,[s_i(l+\ell)]\,\mathrm{V}\,[s_j(l)]]^{1/2}}.$$

If the individual processes are jointly stationary, the correlation coefficient does not depend on bin index l and this fact can be used to incorporate more data into the estimation process. If the correlation coefficient is zero for all lags ℓ and bin indices l, the digitized processes are uncorrelated *and*, in addition, when the number of counts in a bin is binary valued, they are also statistically independent. But note that when only binary values occur, the statistical description for jointly Bernoulli processes [18] requires the correlation coefficient to be greater than $-[p_i p_j/(1-p_i)(1-p_j)]^{1/2}$, where $p_i = \mathrm{E}[s_i]$. Since $p_i \approx \lambda \cdot \delta$, the most negative a correlation coefficient can be depends on the binwidth δ. Thus, if negative correlation is measured, the actual correlation between the point processes might be more negative.

The usual way of estimating the pairwise correlation coefficient is to use the so-called plug-in estimator: use empirical estimates for each expected value and variance in the formula. In the Bernoulli case, the variance depends only on the probability of an event occurring to – $\mathrm{V}[X] = \mathrm{E}[X](1-\mathrm{E}[X])$ – thereby simplifying the estimation process. No empirical estimate will be zero; consequently, confidence intervals need to be established to test for independence.

Even if all $\binom{M}{2}$ pairs of processes in a collection of M are statistically independent, the processes could be statistically dependent. To prove that a collection of jointly Bernoulli processes consists of statistically independent components, the correlation coefficient of all orders would need to be zero. Defining a high-order correlation coefficient consistent with the pairwise correlation is not obvious. The one definition that can be used for all orders and has the property that all correlation coefficients are less than 1 can be derived [15]. The formula for the kth order correlation coefficient function

$$\varrho_{i_1 i_2 \ldots i_k}(l, \ell_1, \ldots, \ell_{k-1}) = \frac{\mathrm{E}[(s_{i_1}(l+\ell_{i_1}) - \mathrm{E}[s_{i_1}(l+\ell_{i_1})]) \cdots (s_{i_k}(l) - \mathrm{E}[s_{i_k}(l)])]}{(\mathrm{V}[s_{i_1}(l+\ell_{i_1})] \cdots \mathrm{V}[s_{i_k}(l)])^{1/k}},$$

which clearly is a complicated function of k variables. Furthermore, the amount of data needed to compute it grows exponentially in k; rarely is enough data available to reliably estimate a correlation function of orders greater than 2. Furthermore, the correlation coefficients of various orders are loosely related to each other [14]. For example, for a completely symmetric population (all selections of k neurons are correlated in the same way), if the pairwise correlation coefficient exceeds $1/M$, then the third-order correlation coefficient must be nonzero. Note that the critical quantity here is *not* the number of neurons recorded but the actual population size. Consequently, the correlation coefficients of all orders may be linked in ways that are not apparent from measurements. Other techniques for testing for dependence among k processes based on entropy calculations promise to be less data hungry, but it remains an open problem to assess dependencies at different lags.

4.4
Dynamic Neural Processing

All of the techniques described previously assume stationarity (estimates of the interval histogram and correlation coefficients) or periodic repetitions of the stimulus (the PST histogram). Even under these restrictions, much insight can be gained into the nature of the point process's intensity. Can this information be exploited to determine the neural response, even interpret it, under less well-controlled conditions, for example, in freely behaving conditions? Using dynamic Bayesian models and adaptive filters, such processing possibilities can be derived [8, 19, 20].

For any dynamic system that begins functioning at the time origin, the input–output relationship is given by the conditional probability function $p_{Y_t|x_{0:t}}(y_t|x_{0:t})$, where $x_{0:t}$ denotes the entire input over the interval $[0, t]$ and y_t is the output at time t. Note that the system could introduce stochastic behavior on its own. Foreshadowing binned recordings of neural spike trains, we assume the time variable is integer valued. In many cases, this general form can be expressed in a recursive form reminiscent of a state space model.

$$p_{Y_t|x_{0:t}}(y_t|x_{0:t}) = p_{Y_t|x_t,\, y_{0:t-1}}(y_t|x_t, y_{0:t-1}).$$

In words, knowledge of the previous outputs and the most recent input suffice to determine the statistical properties of the output at each moment of time. Note that because of the form of the intensity, the probability law of point processes (4.1) has the same form.

The goal is to estimate the signal driving the intensity based on observing the response of one or several neurons. A well-known result from estimation theory states that the minimum mean-square estimate of the input x_t based on observing the output $y_{0:t}$ for *any* system is the conditional mean: $\hat{x}_t^{\text{opt}} = \text{E}[X_t|Y_{0:t} = y_{0:t}]$. By using this estimate, you can determine the system's input knowing the output signal and the system's input–output relationship as summarized by $p_{Y_t|x_{0:t}}(y_t|x_{0:t})$. In neural processing applications, given recorded spike trains and some knowledge of the neuron's intrinsic properties, you can estimate the stimulus being encoded by the spike train(s). To find it, you need the probability function $p_{X_t|y_{0:t}}(x_t|y_{0:t})$, known as the posterior distribution. The point process model gives us the opposite conditional probability. To convert from the model to the posterior, we use Bayes' rule and some simplifications.

$$\begin{aligned} p_{X_t|Y_{0:t}}(x_t|y_{0:t}) &= \frac{p_{X_t,\, Y_{0:t}}(x_t, y_{0:t})}{p_{Y_{0:t}}(y_{0:t})} \\ &= \frac{p_{X_t|Y_{0:t-1}}(x_t|y_{0:t-1})\, p_{Y_t|x_t,\, Y_{0:t-1}}(y_t|x_t, y_{0:t-1})}{p_{y_t|y_{0:t-1}}(y_t|y_{0:t-1})}. \end{aligned} \tag{4.9}$$

This result amounts to a nonlinear recursive equation for the evolution of the posterior as observations accumulate, what amounts to a state space model [21].

From this basic result, many algorithms have been developed and applied to the analysis of population data. From multielectrode data, how hippocampal place cells encode spatial information was determined in real time [19, 22]. Receptive field plasticity for these place cells was determined using an adaptive filter strategy [8]. Particle filtering, a sequential Monte Carlo technique for estimating the conditional mean, has also been used to measure these receptive fields [23]. Learning, a situation where a neuron continually changes its characteristics, has been explored by estimating model parameters using the expectation maximization (EM) algorithm [24].

These are just examples of how modern statistical signal processing methods can be used to “read” the neural code given an intensity description. But how well can such methods succeed? Even though these algorithms all have the goal of minimizing an error criterion, what is the smallest achievable error? How many neural recordings need to be made to infer the encoded input to a specific accuracy? A more general theory of information processing – information theory – needs to be explored that sheds light on these questions.

4.5
Information Theory and Neural Signal Processing

In sensory systems especially and in neural information processing systems more generally, we are ultimately concerned with how a neuron’s discharge pattern represents information and what fidelity that representation provides. However, how data are processed affects the quality of the analysis and what we can glean from measurements. Information theory has much to say about what interfaces to neural systems, for both stimulation and recording, can be effective.

The fundamental model for information theory analysis is shown in Figure 4.2. A source produces S, which can be taken to be a stimulus, is encoded into the signal X. The channel introduces noise and possibly distorts the signal to produce the signal Y. The decoder attempts to reconstruct from Y what the stimulus was by producing the estimate $\hat{S}$. More generally, this could be a cascade of arbitrary systems. For example, the channel could be a measurement system trying to capture the encoded stimulus and the decoder could be a signal processing system that extracts information from measurements. In another scenario, the encoder could be a stimulation interface in which the signal X is the result of neural stimulation.

The fundamental quantity of information theory is mutual information. If x and y represent two random variables, the mutual information between them is defined to be

$$I(X;Y) = \iint p_{X,Y}(x,y) \log_2 \frac{p_{X,Y}(x,y)}{p_X(x)p_Y(y)}\, dx\, dy. \tag{4.10}$$

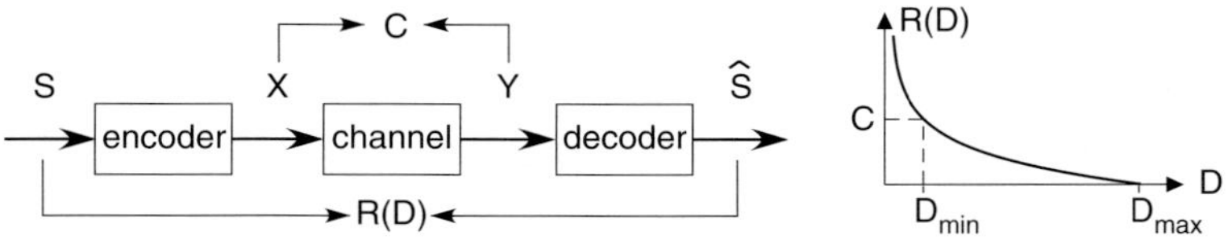

Figure 4.2 The key quantities in Shannon’s classic information theory – capacity and the rate distortion function – are defined for the standard communications and processing model used in information theory. Capacity (C) summarizes the channel that presumably introduces disturbances into the communication process. The rate distortion function $\mathcal{R}(D)$ depends solely on stimulus characteristics. Shannon’s rate distortion theorem relates these two quantities, showing that the smallest possible distortion D_{min} is determined by $C = \mathcal{R}(D_{min})$.

If the random variables are discrete valued, the integrals become sums without affecting any of the properties of mutual information. The use of a base-two logarithm means the "units" of mutual information are bits. Two important properties of mutual information are (1) that $I(X; Y) = 0$ – mutual information must be nonnegative – only when X and Y are statistically independent and (2) that its maximal value is achieved when $X = Y : I(X; X) = H(X)$ in the discrete-valued case and $I(X; X) = +\infty$ in the continuous-valued case. Mutual information completely characterizes the degree of similarity between the statistical properties of two random variables, making it a far more powerful measure of statistical dependence than the correlation coefficient because these properties apply regardless of the random variables involved. If X is a system's input and Y is its output, zero mutual information means the output does not depend on the input and maximal mutual information means the output echoes the input exactly. More generally, the smaller the mutual information relative to its maximum, the less Y represents X. In the communications scenario, the mutual information between the channel's input and the output is significantly smaller than its maximum because of disturbances introduced by the channel.

4.5.1
Data Processing Theorem

One of the most important post-Shannon results in information theory, the data processing theorem, highlights the fragility with which signals represent information. Using Figure 4.2 as an example, consider two systems of any kind – linear, nonlinear, time invariant, time varying – in cascade. We use the notation $X \to Y \to Z$ to indicate the signals involved. The *data processing theorem* is quite succinctly stated.

$$I(X; Y) \geq I(X; Z).$$

First of all, this result means that regardless what the second system is, the degree to which its output Z reflects the input X cannot be better than the representation by Y. In fact, unless the system is very carefully chosen, the representation of X by Z will be worse! For example, the point process digitization process most certainly reduces the capability of extracting information about the original point process. However, putting a quantitative measure on the reduction and interpreting it is quite difficult, partially because calculation of mutual information in this case is very difficult. But more importantly, applying the data processing theorem in this form demands a cascade of two systems. The data processing theorem can be shown to hold between a single system's input and output using a different information theoretic measure [15, 25, 26]. Consequently, interfacing to the brain will probably, but not always, lead to performance reductions because of this result. We need more details than this; Shannon theory tilted toward the general communications and processing scenario (not just digital communications) suffices to some degree. We need two quantities to delve more deeply into information process: the channel capacity and the rate distortion function.

4.5.2 Channel Capacity

The *capacity* defines the ultimate ability to extract information about the channel's input from its output [27]. For a point process, we take the intensity as the channel input since it expresses how the information (the stimulus or the motor command) is encoded by the point process. The output is, of course, the actual sequence of events (the spike train) produced according to this model. More generally, when a measurement system interacts with a system under study, capacity determines how accurate the measurements can be. Mutual information expresses the statistical dependence of the output on the input, but it depends on the choice for the input distribution. To obtain a quantity that expresses the best possible situation, Shannon optimized mutual information over all input probability distributions. Maximizing in this way yields the *channel capacity*, a fundamental quantity that characterizes *any* channel.

$$C = \lim_{T\to\infty} \max_{p_X(\cdot)\in\mathcal{C}} \frac{1}{T} I(X;Y). \tag{4.11}$$

As mutual information usually increases in proportion to the length of the observation interval T, we divide by it, which results in capacity having units of bits/s. The maximization is usually restricted to probability distributions for the channel input that has characteristics defined by the constraint class $\mathcal{C}$. The oft-quoted channel example is an additive Gaussian noise channel wherein $Y = X + N$, N being Gaussian noise. The constraint on the channel input is a mild one: demand that the input power $\mathrm{E}[X^2]$ be less than a specified value P. Shannon showed that the capacity was achieved when the input is Gaussian and that the capacity equals $\frac{1}{2}\log\left(1+\frac{P}{\sigma_N^2}\right)$.

Kabanov [28] derived the capacity of the single point process channel, and Johnson and Goodman [29] extended his result to the multiple Poisson process case. Both results impose minimal and maximal constraints on the intensity: $\mathcal{C} = \{\mu(t;N_t,\mathbf{w}_t) : \lambda_{\min} \le \mu(t;N_t,\mathbf{w}_t) \le \lambda_{\max}\}$. If the maximal rate were not constrained, the capacity would be infinite. Note that this maximal rate equals the peak rate a given neuron can produce, even though it may be capable of doing so only transiently. Kabanov found that $C^{(1)}$, the capacity of the single neuron channel, to be attained by a Poisson process and was related to the constraints according to

$$C^{(1)} = \frac{\lambda_{\min}}{\ln 2}\left[\frac{1}{e}\left(\frac{\lambda_{\max}}{\lambda_{\min}}\right)^{\lambda_{\max}/(\lambda_{\max}-\lambda_{\min})} - \ln\left(\frac{\lambda_{\max}}{\lambda_{\min}}\right)^{\lambda_{\max}/(\lambda_{\max}-\lambda_{\min})}\right].$$

Typically, a neuron's rate of discharge in response to stimuli can fall to zero, making $\lambda_{\min} = 0$. In this case, the expression for capacity simplifies greatly and we shall frequently use this result in subsequent expressions.

$$C^{(1)} = \frac{\lambda_{\max}}{e\ln 2}. \tag{4.12}$$

Kabanov further showed that no non-Poisson process satisfying the constraints can have a larger capacity. Our recent work [15] details this result, showing that more realistic point process models that embrace refractory effects and other dependencies on previous events have a strictly smaller capacity neatly given by the expression

$$C^{(1)} = \max_{\mu(t;N_t,\mathbf{w}_t)\in\mathcal{C}} \frac{\mathrm{E}[\mu(t;N_t,\mathbf{w}_t)]}{e\ln 2},$$

where $\mathrm{E}\left[\mu(t;N_t,\mathbf{w}_t)\right]$ equals the expected value of intensity with respect to all possible histories. This expected value is exactly the same as the one that appears in (4.8), the expected value of the PST histogram. For a Poisson process that has no history effects, this quantity equals $\lambda_{\max}$. For more realistic models of single-neuron discharge characteristics, the average rate is strictly less than $\lambda_{\max}$, thereby resulting in a smaller capacity. For example, if absolute refractory effects occur (refractory interval Δ), the capacity under a maximal rate constraint of $\lambda_{\max}$ is given by

$$C^{(1)} = \frac{1}{1+\lambda_{\max}\Delta}\cdot\frac{\lambda_{\max}}{e\ln 2}. \tag{4.13}$$

Capacity results are quite limited for neural populations. In contrast with Kabanov's general single point process result, multineuron results are confined to jointly *Poisson* process descriptions of a restricted type [14]. Two issues arise in an information theoretic analysis of a population's capacity. The first is the input innervation: do the neurons comprising the population have individual innervation, do they share a single input, or do they have a combination of the two? Individual innervation is shown in Figure 4.3a. From an information theoretic perspective, the separate signals arise from encoding the stimulus into M signals in a coordinated fashion. The decoder must use all the outputs (spike trains) to decode as much as possible. The single-input scenario is shown in Figure 4.3b, wherein the encoder produces one signal that serves as the common intensity of all the point process

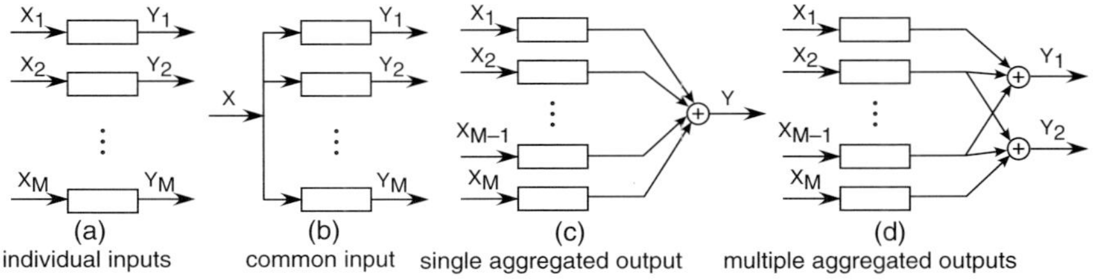

Figure 4.3 The panels show simple interfacing scenarios to neural systems. In panel (a), each neuron receives its own input, while in (b) a single input drives all. In panels (c) and (d), the neural outputs are summed, first with only one composite output and the rightmost with several outputs that aggregate overlapping subpopulations.

channels. Second, are there lateral connections among the population's members? In Figure 4.3, the neurons function independently, with the occurrence of a spike in one not affecting the production of a spike in another. In this case, conditioned on the intensity, the point processes are independent and have zero correlation coefficient values (if the time variations in X are taken into account). If lateral connections do exist, at least some of these correlations would be nonzero. Interestingly, when no lateral connections exist (zero correlation), input innervation (at least for the two extremes shown in Figure 4.3) does *not* affect the capacity result. Consequently, the capacity of the channel does not depend on which of these two ways of interfacing to the subchannels is used. For a population of M neurons, each of which is modeled as a Poisson process having a maximal firing rate of λ_{max}, the population capacity $C^{(M)}$ in this case simply equals $MC^{(1)}$. Thus, the capacity is proportional to the population size, which makes a population far more capable of encoding a stimulus to a high degree of fidelity than a single neuron.

When lateral connections among population members do exist, capacity depends on the degree of connection-induced correlation and on the input innervation. If a common input serves the population, increased coupling between neurons results in a capacity *smaller* than $mC^{(1)}$. On the other hand, if each neuron has its own input, a different result emerges: capacity increases with increased coupling, attaining values that can double it from the uncoupled baseline. In either case, the presumption is that the encoder and decoder know the channel's characteristics and function optimally to cope with the channel's behavior.

4.5.3
Rate Distortion Theory

Shannon's information theory goes well beyond dealing solely with communication systems and shows that by using the capacity how the channel – any system in fact – affects recovering encoded information from the channel's output is *entirely* summarized by the capacity. Letting S represent the stimulus and $\hat{S}$ its "estimate" after it has been encoded, passed through a channel, and processed (Figure 4.2), this theory begins by introducing a *distortion measure* $d(S, \hat{S})$ that expresses how the fidelity of the communication (or signal processing) system is to be assessed. Presumably, the distortion increases as the discrepancy increases between the stimulus and its reconstructed value. Interestingly, Shannon's framework allows *any* reasonable distortion measure ($d(S, \hat{S}) \geq 0$, equaling zero when $\hat{S} = S$). It can be chosen according to whatever criteria evaluate what effective communication and processing mean in any particular scenario. A common distortion measure used in communications and signal processing is the squared error measure: $d(S, \hat{S}) = (\hat{S} - S)^2$. The distortion measure could also incorporate some desired stimulus processing, in which $\hat{S}$ relates to some feature of S. For control applications, the distortion could incorporate time-to-target penalties.

Next, Shannon defined the average distortion $\overline{D}$ as the expected value of the distortion measure with respect to the joint distribution of the stimulus and its estimate.

$$\overline{D} \triangleq \mathrm{E}[d(s,\hat{s})] = \iint d(s,\hat{s}) p_{\hat{s}|s}(\hat{s}|s) p_s(s)\, ds\, d\hat{s}.$$

The conditional distribution $p_{\hat{S}|S}(\hat{s}|s)$ depends on virtually everything in a neural coding scenario: how the stimulus is encoded, the neuron's spiking characteristics, and how the decoder works. Recall that neuroscientists rarely know (or can measure) $\hat{S}$, what is presumably the estimated stimulus. In biohybrid applications, however, particularly those involving an effector, estimation error (distortion) can be explicitly measured. Shannon proceeded by defining the *rate distortion function* $\mathcal{R}(D)$ to be the minimal mutual information between the stimulus and its estimate over *all* possible channels, encoders, and decoders that yield an average distortion smaller than D.

$$\mathcal{R}(D) \triangleq \lim_{T\to\infty} \frac{1}{T} \min_{p_{\hat{S}|S}(\cdot|\cdot):\bar{D}\le d} I(S;\hat{S}). \tag{4.14}$$

Note that the minimization is calculated over *all* possible relationships between a stimulus and its estimate, not just the one under study. In this way, we do not need to specify the optimal estimator. The rate distortion function has units of bits/s. It is *not* the bit rate of some equivalent digital scheme; rather it becomes the intermediate variable, a universal unit of information, that evaluates how well communication and processing systems can perform according to the specified distortion criterion.

Rate distortion functions are notoriously difficult to calculate, with only a few results known. If the stimulus source is a band-limited Gaussian random process having maximal power P and bandwidth W, and the distortion measure is squared error, the rate distortion function equals [27]

$$\mathcal{R}(D) = \begin{cases} W \log_2 \dfrac{P}{d}, & D \le P, \\ 0, & D > P. \end{cases} \tag{4.15}$$

This result, shown in Figure 4.2, illustrates the properties *all* rate distortion functions satisfy.

- $\mathcal{R}(D)$ is a strictly decreasing and convex function.
- $\mathcal{R}(D)$ equals zero at some distortion, remaining zero for any larger values. This critical value is known as the maximal distortion $d_{\max}$ and corresponds to the decoder's best guess as to what the stimulus might be with *no* data. For example, in the Gaussian case, mean-squared distortion is minimized by guessing $\hat{S} = \mathrm{E}[S]$. In the binary case, the minimum probability of error decoder flips a coin biased to the stimulus probability p. Consequently, no combination of encoder, channel, and decoder should yield a distortion larger than $d_{\max}$.

Shannon's crowning result, the rate distortion theorem, unifies all of his results: The distortion at which the rate distortion function equals the channel capacity defines the smallest possible distortion $D_{\min}$ *any* encoder and decoder can obtain: $\mathcal{R}(D_{\min}) = C$ (see Figure 4.2). The rate distortion function depends only on stimulus characteristics and on the desired level of distortion. Capacity summarizes the interference properties any system has that intervenes between the encoder and the decoder. The capacity determines via the rate distortion function the smallest possible distortion that any encoder and decoder can achieve for a given source. The value of $D_{\min}$ defined by the source and the channel determines how well a given system can perform, thereby serving as a benchmark. In more detail, Shannon showed there exists an encoder and decoder that can achieve this performance limit. Unfortunately, his theory does not reveal what they are; it merely states they exist.

It is important to appreciate how important and general Shannon's result is. Rate distortion theory applies when *no* digital scheme is involved. Quite surprisingly, "bits/s" is the fundamental unit of exchange in *any* communication or signal processing system, whether it communicates discrete- or continuous-valued signals and regardless of the criteria used to judge performance (the distortion measure). Information theoretic rate (not to be confused with spike rate) serves as an intermediary. Regardless of the distortion measure, *all* rate distortion functions look like the one portrayed in Figure 4.2. Second, an important, general concept emerges that highlights the importance of capacity: the larger the capacity the smaller the achievable distortion can be in *all* cases.

4.5.4 Application to Biohybrid Systems

To show the relevance of rate distortion theory to biohybrid systems, we return to the population capacity results. Because the capacity is proportional to the population size, the smallest possible distortion must decrease accordingly. The nature of the decrease depends on the details of the rate distortion function. Using the Gaussian result of (4.15) as an example, we find that $D_{\min} = \exp\{-MC^{(1)}\ln 2/W\}$: the distortion decreases *exponentially* in a number of neurons. Consequently, with a suitable encoder and decoder, a neural population can represent information many times better than can a single neuron. Furthermore, the proper choice of interconnections among the neurons can increase the capacity, making a population of a given size even more capable of supporting a low-distortion output.

Note that when the neurons function independently, capacity remained unchanged when the encoder provided the same or individual signals to the neural subchannels. When the subchannels were conditionally (conditional on the input) statistically dependent, the capacity changed in these two conditions. This result has important implications for stimulation-type interfaces. Presumably, electrical stimulation will affect several neurons, essentially serving as a common input. If we take individual innervation as the baseline, imposing a common input can affect the capacity. If the neurons function independently, *no* capacity change will occur, meaning that artificial stimulation can function as well as the natural one. The utility of this result hinges on finding an effective encoder and the brain adapting to the new information representation. On the other hand, if the neurons coordinate their activity, using electrical stimulation can lower the capacity. In the context of rate distortion theory, the smaller capacity means the minimal distortion $D_{\min}$ must increase, resulting in a performance deficit that cannot be overcome.

Electrophysiological techniques for recording neural signals, either for scientific study or for controlling prostheses, employ electrodes that each sum the electric fields produced by several neurons. For single-neuron recordings, electrode position is adjusted so that one spike train dominates and is well isolated. In such circumstances, the only confounding effects are recording noise and background gross potentials. Johnson [17] showed that recording noise acts as a low-pass filter, the cutoff frequency of which equals the signal-to-noise ratio times the rise time of each spike, and that background signals can affect the measured point process intensity according to their derivative. When the spike train is not well isolated, whether it be from a single or multielectrode recording, either spike sorting is used to convert recordings into spike trains, each of which supposedly represents the activity of a single neuron, or the recordings are used "as is" in simplistic prosthetic applications. In either case, how well the result represents the neural activity depends on the algorithm used. Details vary from algorithm to algorithm, but information theory places a hard limit on the fidelity they can achieve.

First of all, we know from the data processing theorem that once summed by the recording electrode, no algorithm that follows can increase fidelity. Johnson and Goodman [29] examined the consequence on the population capacity of summing the activity of the entire population or summing the activities of several overlapping subpopulations (see Figure 4.3). In the first instance wherein only one electrode records the activity of many neurons, the channel capacity is $1.577C^{(1)}$, a little over one-and-a-half times the capacity of one neuron.[2)] Even though the best possible spike-sorting algorithm may follow, the total capacity cannot exceed this small value. If E electrodes are used, the best possible situation is for each to record from half the population (but different subpopulations), in which case the capacity becomes $(2E-1)\,1.577C^{(1)}$: the capacity is proportional to the number of electrodes in the array. How to translate these results into a measure of prosthetic errors is difficult. Suffice it to say, the smaller the capacity, the larger the best possible distortion that any system can incur when it tries to extract information about the source. In summary, regardless of an algorithm's sophistication in decoding neural signals, measurement prostheses cannot penetrate the performance barrier established by the rate distortion theorem. This result suggests that recording techniques must improve so that the activity of large numbers of neurons can be measured before *any* measurement prosthetic algorithm can produce high-fidelity actions.

4.6 Summary

Point processes, the ubiquitous model of neural discharge patterns, are completely characterized by their intensities. Neural signal processing for the scientist amounts to estimating the intensity, not only for individual neurons but also for the joint

2) This result assumes that spike waveforms are identical. If the waveforms or even just their amplitudes differ, this result is pessimistic.

response of a neural population. Partially because no convenient model for jointly distributed point processes exist, how to characterize a joint response is an open research issue. The most direct approach is to measure PST histograms for each neuron and to compute pairwise correlation functions. However, this approach does *not* suffice to characterize fully the population's response. More promising are the dynamic, state-space-like approaches, which are very flexible but often computationally intensive.

The introduction of information theory, specifically rate distortion theory, into signal processing helps frame the performance limits any algorithm can achieve. The theory's generality is both useful and frustrating. Modern methods for recording electric fields produced by neurons cannot overcome the capacity limits the aggregation of individual responses into a superposition imposes. Exactly what these are is the frustrating part: information theory only reveals that no matter what the distortion measure used, aggregation reduces capacity thereby increasing error.

More telling is the prediction that stimulation interfaces can work very well. Information theory throws up no roadblocks here like it does for gross recordings. However, this result comes with the admonishment that a well-suited encoder must be found (an engineering design problem) and that the decoder, the brain, can adapt to the new way information is represented. The same holds for recording techniques that produce individual spike trains, whether it be by careful electrode position or some nonelectrical technique: no capacity loss impairs optimal performance but the encoder – the nervous system – must adapt to the decoder, the apparatus designed to translate neural signals into action.

References

1 Snyder, D.L. (1975) *Random Point Processes*, John Wiley & Sons, Inc., New York.

2 Daley, D.J. and Vere-Jones, D. (1988) *An Introduction to the Theory of Point Processes*, Springer, New York.

3 Johnson, D.H. (1996) Point process models of single-neuron discharges. *J. Comput. Neurosci.*, **3**, 275–299.

4 Cox, D.R. (1962) *Renewal Theory*, Methuen, London.

5 Johnson, D.H. and Swami, A. (1983) The transmission of signals by auditory-nerve fiber discharge patterns. *J. Acoust. Soc. Am.*, **74**, 493–501.

6 Miller, M.I. (1985) Algorithms for removing recovery-related distortion from auditory-nerve discharge patterns. *J. Acoust. Soc. Am.*, **77**, 1452–1464; Also see Erratum. (1986) *J. Acoust. Soc. Am.*, **79**, 570.

7 Hawkes, A.G. (1971) Spectra of some self-exciting mutually exciting point processes. *Biometrika*, **58**, 83–90.

8 Brown, E.N., Nguyen, D.P., Frank, L.M., Wilson, M.A., and Solo, V. (2001) An analysis of neural field plasticity by point process adaptive filtering. *Proc. Nat. Acad. Sci. USA*, **98**, 12261–12266.

9 Paninski, L. (2004) Maximum likelihood estimation of cascade point-process neural encoding models. *Network: Comp. Neural Syst.*, **15**, 243–262.

10 Nicolelis, M.A.L. (ed.) (1999) Methods for neural ensemble recordings, in *Methods & New Frontiers in Neuroscience*, CRC Press, Boca Raton, Florida.

11 Sheikh, M.A. (2006) Fundamental limits in spike sorting. Master's thesis, Rice University, Houston, Texas.

12 Gozani, S.N. and Miller, J.P. (1994) Optimal discrimination and classification of neuronal action potential waveforms

from multi-unit, multi-channel recordings using software-based linear filters. *IEEE Trans. Biomed. Eng.*, **41**, 358–372.

13 Lewicki, M.S. (1998) A review of methods for spike sorting: the detection and classification of neural action potentials. *Network: Comp. Neural Syst.*, **90** (4), R53–R78.

14 Johnson, D.H. and Goodman, I.N. (2009) Jointly Poisson processes. arXiv:0911.2524.

15 Johnson, D.H. (2010) Information theory and neural information processing. *IEEE Trans. Info. Th.*, **96** (2), 653–666.

16 Johnson, D.H. (2008) The correlation function of multiple dependent Poisson processes generated by the alternating renewal process method. arXiv:0811.3366.

17 Johnson, D.H. (1978) The relationship of post-stimulus time and interval histograms to the timing characteristics of spike trains. *Biophys. J.*, **22**, 413–430.

18 Johnson, D.H. (2004) Neural population structures and consequences for neural coding. *J. Comput. Neurosci.*, **16**, 69–80.

19 Brown, E.N., Frank, L.M., Tang, D., Quirk, M.C., and Wilson, M.A. (1998) A statistical paradigm for neural spike train decoding applied to position prediction from ensemble firing patterns of rat hippocampal place cells. *J. Neurosci.*, **18**, 7411–7425.

20 Smith, A.C. and Brown, E.N. (2003) Estimating a state-space model from point process observations. *Neural Comput.*, **15**, 965–991.

21 Paninski, L., Ahmadian, Y., Ferriera, D., Koyama, S., Rahnama, K., Widne, M., Vogelstein, J., and Wu, W. (2010) A new look at state-space models for neural data. *J. Comput. Neurosci.*, **29**, 107–126.

22 Barbieri, R., Frank, L.M., Nguyen, D.P., Quirk, M.C., Solo, V., Wilson, M.A., and Brown, E.N. (2004) Dynamic analyses of information encoding by neural ensembles. *Neural Comput.*, **16**, 277–307.

23 Ergun, A., Barbieri, R., Eden, U.T., Wilson, M.A., and Brown, E.N. (2007) Construction of point process adaptive filter algorithms for neural systems using sequential Monte Carlo methods. *IEEE Trans. Biomed. Eng.*, **54**, 419–428.

24 Wirth, S., Yanike, M., Frank, L.M., Smith, A.C., Brown, E.N., and Suzuki, W.A. (2003) Single neurons in the monkey hippocampus and learning of new associations. *Science*, **300**, 1578–1584.

25 Sinanović, S. and Johnson, D.H. (2007) Toward a theory of information processing. *Signal Process.*, **87**, 1326–1344.

26 Johnson, D.H., Gruner, C.M., Baggerly, K., and Seshagiri, C. (2001) Information-theoretic analysis of neural coding. *J. Comput. Neurosci.*, **10**, 47–69.

27 Cover, T.M. and Thomas, J.A. (2006) *Elements of Information Theory*, 2nd edn, John Wiley & Sons, Inc., New York.

28 Kabanov, Y.M. (1978) The capacity of a channel of the Poisson type. *Theory Prob. Appl.*, **23**, 143–147.

29 Johnson, D.H. and Goodman, I.N. (2008) Inferring the capacity of the vector Poisson channel with a Bernoulli model. *Network: Comp. Neural Syst.*, **19**, 13–33.

5
Dynamic Clamp in Biomimetic and Biohybrid Living-Hardware Systems

Ryan Hooper and Astrid A. Prinz

The dynamic clamp is an experimental technique that enables the selective introduction of artificial conductances in electrically excitable cells. This is accomplished by establishing a hybrid system composed of biological tissue and a controlling computational device. This basic ability endows the method with not only a flexibility for probing a wide variety of preparations but also a wide variety of potential uses for study of the dynamics of excitable tissue. Unlike a number of the other hybrid systems covered in this book, the dynamic clamp is primarily employed as a basic research tool, with no therapeutic applications developed using it to date.

In this chapter, we will describe how the dynamic clamp creates an artificial conductance, review some of the implementations in use, explore a few examples of what kind of experiments are possible with this technique, and consider some of the recent discoveries of the dynamic clamp's limitations and future potential.

5.1
What is a Dynamic Clamp?

The dynamic clamp, also known as the conductance clamp, is a powerful tool developed for studying the electrophysiological behavior of cells. It is composed of a computational device connected to an electrically excitable cell or network of cells. In many respects, the technique is very familiar: an electrically excitable cell is impaled with an electrode that serves as an interface for the experiment, which is recorded from and controlled by a user-operated tool that serves to observe and manipulate the electrical dynamics of the cell. This reminds us of the voltage clamp and current clamp, each of which independently holds its designated electrical property (either voltage or current) in the cell fixed (i.e., "clamped") for a given time period, while simultaneously recording the other (current or voltage, respectively). While these longer established electrophysiological tools are primarily geared toward manipulating cellular electrical properties in order to measure cellular activity dynamics, the dynamic clamp's unique manipulation allows it to actually alter the behavior of cells by perturbing the very operating principle upon which its electrical

Biohybrid Systems: Nerves, Interfaces, and Machines, First Edition. Edited by Ranu Jung.

activity ultimately relies: conductance. This transmembrane conductance is the basis of electrical activity in electrically excitable cells, be it via electrical cell-to-cell coupling, voltage-dependent ion channels, or voltage-independent ion channels (such as those controlled by synapses, giving rise to synaptic currents) [1]. The dynamic clamp introduces artificial conductances that either act in parallel with or in lieu of such normally existing conductances.

In a sense, the dynamic clamp can be thought of as a current clamp that is controlled by a feedback loop that determines the appropriate current to use at any instant (Figure 5.1). Unlike traditional current clamp, the current to be used is not known *a priori*. Instead, the current necessary to establish a given artificial conductance is calculated using the recorded membrane potential. This computation relies on Ohm's law, $I = g^* V$, where I is the membrane current to be injected into the cell in order to adjust the effective membrane conductance, g is the target artificial membrane conductance, and V is the driving force of the biological component being simulated. The final expressions for driving force and conductance are chosen to simulate the dynamical properties of the biological component to be modeled. Typically, this current must be calculated very quickly to keep up with the dynamics of the system being interfaced with.

This process is repeated cyclically based on a feedback principle. During any given feedback loop, the device cycles between recording membrane voltage, making a calculation of needed current, and injecting current. The calculation will typically depend on membrane voltage, and can be thought of as a computational model of the biological component that the experimentalist intends to mimic: whether it be a gap junction, voltage-dependent ion channel, synapse, model cell, or even model network of cells. What is important is that the process to be modeled must have dynamics and topology that can be approximated with a computational model and that those dynamics and topology be programmable within the framework of the device (whether it be software or hardware) used to implement dynamic clamp, and that

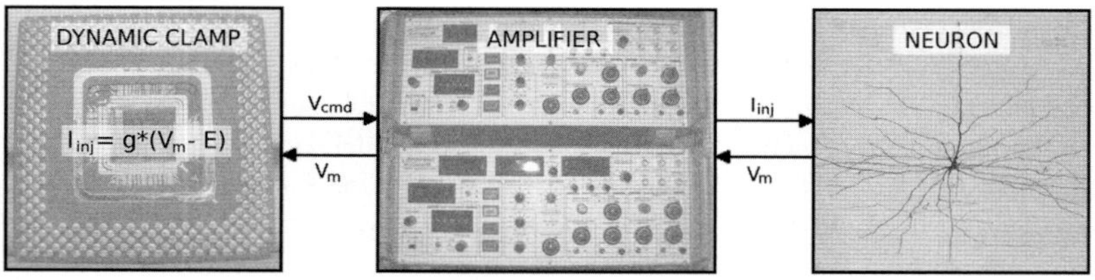

Figure 5.1 Typical dynamic clamp feedback loop. The system cycles between recording the neuron's membrane voltage (V_m), then calculating and applying a current (I_{inj}) to impose a calculated conductance change (g) at a reversal potential (E). Typically, these operations are performed with the same electrode, which necessitates a time-sharing scheme. Note the intermediary presence of the electrophysiology amplifier. This provides the time-sharing scheme, translates the dynamic clamp's representation for I_{inj} from a command voltage (V_{cmd}) into a current, and performs other required functions to record from the electrode. Performance limitations of the amplifier can make it a source of error in dynamic clamp experiments [22].

the complexity of the model works within the performance constraints of the dynamic clamp system.

One of the simplest operations that the dynamic clamp performs is that of electrical coupling, which is an equivalent computation to introducing a fixed conductance between two cells. This was the operation performed when the dynamic clamp was first introduced [2–5], when isolated embryonic chick ventricular cells were coupled with an artificial electrical junction of a selected intercellular conductance, originally termed an "Ersatz Nexus." In this case, the cells were physically connected by a dynamic clamp that consisted of analog circuitry that established a resistive connection between the cells. This was accomplished by a difference computation on the independently measured membrane potentials of the two cells, multiplying this difference (driving force) by the desired conductance to obtain the junction current, and injecting the resulting current into each cell, with opposite polarities, so that the current may flow from one cell to the other, thus providing each cell with input similar to what it would receive if it were coupled to the other via a real gap junction. The resulting operation can then be described as

$$I_1 = G(V_1 - V_2),$$
$$I_2 = G(V_2 - V_1),$$

where V_i and I_i are the respective measured membrane voltages and calculated injection current for each cell, and G is the (usually constant) conductance of the synapse established between them. Note that the magnitudes of I_1 and I_2 are equal.

Other dynamic clamp tasks that go beyond a fixed conductance are performed in a similar fashion, by accounting for the applicable parameters and parameter dependencies in the model. So to simulate voltage-gated ion channels, the expression for conductance is represented by a Hodgkin–Huxley model with attendant activation and inactivation state variables [6, 7], or even a Markov model [8, 9], while the driving force is represented by the difference between the neuron's membrane potential and the reversal potential for that ion channel type. Ligand-gated conductances are created in a similar fashion, where activity for the synaptic input is formulated and used to activate the conductance of a model ligand-gated ion channel. In these two previous cases, our simulated conductance has now evolved to a dynamic rather than a fixed one. It allows for the enhancement of existing ion channels or even the introduction of ion channels that are novel to the preparation. Another dynamic clamp technique recently developed, known as the "dynamic action potential clamp," even allows another biological cell to serve as a vessel for biological ion channels that act the part of the model conductance, which is linked to the cell of interest via the dynamic clamp [10–12].

Once the conductance model is established, it is then implemented in the dynamic clamp. The system performing this execution may operate based on analog, as we saw earlier, or as is becoming most common, digital computational devices, as developed independently by Sharp *et al.*, Robinson and Kawai, and Hutcheon *et al.* [7, 13–15]. As the system runs through its feedback cycle of recording, making calculations, and injecting current, we notice that there are two unique, independent modalities of activity that are interacting with one another: the biological voltage activity in the cell(s) and the model activity in the dynamic clamp. In order to understand how to perform a dynamic clamp task, we must consider the physical interaction of these two systems and their timescales, as the computational system "reads" and responds to the biological system and the biological system "reads" and responds to the computational system.

5.1.1
The Digital Dynamic Clamp

A dynamic clamp implemented with a digital system obtains all of the information about the interfaced biological system through a series of periodically measured voltage *samples*, each of which is simply an instantaneous measurement of the recorded voltage at a given discrete time. This process, in which an analog (in our case biological) waveform is given a digital representation by assignment of an amplitude value and corresponding time value for a periodic series of discrete points over a specified time interval, is known as analog to digital conversion (A/D). The rate at which these samples are acquired by the dynamic clamp is known as the *sampling frequency*, and is typically limited by the performance of the hardware, computational demands of the software, and sometimes the number of simultaneous recording channels chosen to be utilized by the user (as for a computer data acquisition (DAQ) board). At a minimum, we need this sampling frequency to be high enough to track the fastest dynamics present in the activity of the biological system under study. This is quantified by the *Nyquist–Shannon sampling theorem*, which requires us to sample at a rate that exceeds twice that of the fastest dynamics in our sampled activity. However, this performance requirement does not account either for stability considerations that arise from the fact that the dynamic clamp establishes a feedback loop or for imperfections in model implementations and numerical equation solvers, so it is not necessarily a comprehensive performance guideline for achieving satisfactorily performing dynamic clamp in itself. Each sample is also quantized, which is to say that it is given a finite digital value whose resolution is determined by the precision of the system. This is a source of potential measurement error, known as quantization error, so the full scale range of the A/D system should be adjusted to match the amplitude of the biological signal.

Once the digital dynamic clamp system has these samples on hand, it uses them to inform and update the model. The information that will be utilized by the model is determined by its physical dependencies. Sometimes, this may include elements that are not directly measured by the dynamic clamp, such as intracellular calcium concentration, in which case the activity in question must be approximated with a model of its own. Usually the modeled mechanisms are described with ordinary differential equations (ODEs), as is the case for both Hodgkin–Huxley type models and Markov type models, which instead of being implemented with analog circuitry, must be represented and solved in a manner that works within the framework of discrete time computation. A finite-difference approximation for each of the model equations' derivatives is made, which reduces the differential equations to algebraic equations [16]. The equations are now in a form that can be solved stepwise by a given solver, some of the most popular examples being the forward-Euler, exponential-Euler, and Runge–Kutta solvers.

Once the conductance calculation is performed, as in the first step, the biological system obtains all information from the dynamic clamp via the electrode. A digitally implemented dynamic clamp sends a digital representation of the current to be injected into the digital-to-analog converter (D/A), which creates the necessary analog representation that eventually becomes the injection current, possibly subsequent to some analog gain as needed to compensate and allow for the full output range of the D/A to be used. As this injection current flows through the electrode and into the cell, it alters the charge inside the cell, thus effecting a desired overall conductance change. But this change is limited in that it flows from a small interfacing electrode tip that acts as a point current source, and thus may fail to provide adequate space clamp, depending on the electrotonic size and morphology of the cell [10, 17]. This prevents the dynamic clamp from fully mimicking a biological process due to the absence of *in vivo*-like effects of the imposed conductance such as spatially distributed ionic channel density, second-messenger activation, and intracellular calcium concentration modulation. However, this limitation can be an asset when one seeks to investigate the isolated impact of electrical mechanisms, free from the confounding activity of other mechanisms, as was done in a recent study that demonstrated that it is the voltage dependence of L-type Ca^{2+} channels that results in pacemaker activity of dopamine neurons in the substantia nigra, rather than their Ca^{2+} selectivity [18].

In addition, we require simultaneous membrane voltage recording and current injection. Typically, this is accomplished by time-sharing recording and current injection tasks with the discontinuous single electrode voltage clamp (dSEVC) or discontinuous current clamp (DCC) method [19], which permits simultaneous recording and injection with the same electrode. These methods have the added benefit of limiting interactions between the two tasks, which reduces measurement error.

As a digital dynamic clamp cycles through these various tasks, we assess the performance of the system in terms of three measures: time step duration, jitter, and latency. The time step duration is selected by the experimenter and set as a parameter in the dynamic clamp, usually chosen as the limit of stable operation on most modern systems, around 10–50 kHz for computer-based systems, and higher for hardware-based systems. The system will then attempt to execute each dynamic clamp cycle of sampling, calculating, and injecting current within this time step duration. If there are excessive delays or other such imperfections in the involved hardware, the system may fail to meet the targeted duration. We would prefer that the system was guaranteed to never miss such deadlines, and if it is we say that it gives us "hard real-time" performance. Furthermore, we would wish that each of these time step executions would occur on an infinitesimal timescale to maximize accuracy and account for the fastest dynamics in any studied system of interest. But in practice, the time step duration is constrained on its lower bounds by system latency, and jitter can result in time step variability.

Latency is the time required for the system to execute the necessary calculations, overhead tasks, analog and digital conversions, reads and writes, and other time-consuming operations associated with the directly and indirectly required tasks that perform a dynamic clamp cycle between the time that a voltage sample is read and the corresponding output is updated. Such delays are relatively fixed based on the system's architecture and performance.

Jitter is the cycle-to-cycle variability in the time step duration, which is typically quantified by its variance for our purposes. Such timing inaccuracies generally degrade the effective resolution of digital sampling systems and transform the behavior of the time step duration to that of a random variable. Jitter arises from many sources in digital devices, giving rise to jitter on multiple timescales, and like latency, is an inherent product of the system design [20].

5.2
Dynamic Clamp Performance and Limitations

The first dynamic clamp, as implemented by Scott, was constructed from statically configured analog hardware. Due to the difficulty and expense of reconfiguring such a system, its flexibility and power is limited. It was the partnering of the dynamic clamp with the programmable digital device that unleashed the general applicability and flexibility that the technique is capable of. This approach of modeling processes with a computational system can theoretically extend the performance limits of the dynamic clamp to match that of each successive generation of faster, more powerful computational platforms. Indeed, this has been the case so far. We have seen that as computer performance and memory increase, the ability to perform ever more complex and time-sensitive tasks with the dynamic clamp increases.

However, as timescales decrease and candidate technologies on which to base dynamic clamp improve, it becomes important to assess each contender with a rigorous and objective set of evaluation criteria. The accuracy of dynamic clamp systems is our most important consideration if we want to guarantee that the observed spike shapes, firing rate, or any other metric of cellular electrical activity is due to our dynamic clamp manipulation, not just the product of a systematic error in the system. So we would like to know the specific impact of the quantifiable dynamic clamp system performance measures on accuracy such as jitter, latency, step rate, and numerical equation solver employed. This was exactly the focus of a recent study from which a number of interesting conclusions can be drawn [21]. First, while time step, latency, and ODE solver all proved to be important to accuracy, latency was perhaps the most important. But the activity was most impacted by error in the spike shape, not the firing frequency. While jitter is capable of causing such errors, the error it produced was always bounded by the error produced by the worst-case time step when fixed. The authors also were able to infer some basic guidelines for necessary time step durations for Na^+ and K^+ ion channels, and reveal some error trends that may prove useful for diagnosing an error-prone dynamic clamp.

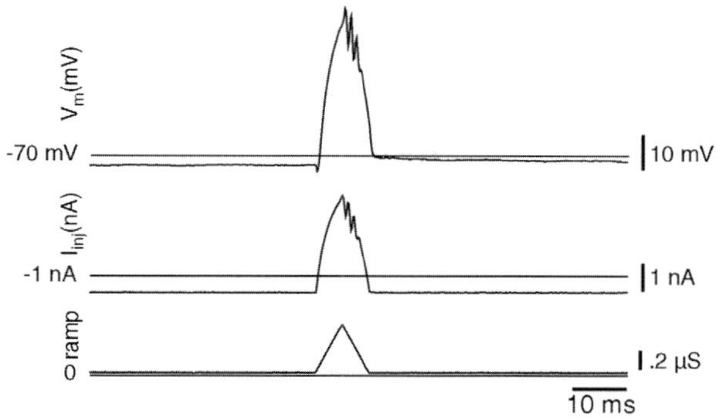

Figure 5.2 Typical dynamic clamp transient instability. A simple up/down conductance ramp introduced into a hyperpolarized neuron in the abdominal ganglion of *Aplysia californica*. The extent of the instability depends on conductance magnitude and exists in most platforms and preparations. Reproduced from Preyer and Butera [22] with permission © 2009 IEEE.

As with any feedback system, instability in the dynamic clamp can lead to system output errors. Preyer and Butera recently quantified the widely observed phenomenon of transient instabilities in voltage levels during large conductance injections using dynamic clamp [22] (Figure 5.2). This phenomenon was verified to be consistently present in a variety of digital dynamic clamp implementations independent of preparation and dynamic clamp system architecture and hardware. An analysis of all of the experimental performance variables that are under the user's control that could have an influence on this phenomenon was performed. The results indicated that a dynamic clamp time step duration only chosen to be fast enough to account for the dynamics of the studied biological system and ensure stability of the ODE solver may not be fast enough to maintain stability of the dynamic clamp. Traditionally, acceptable levels of electrode resistance and capacitance were revealed to be insufficient for optimal dynamic clamp stability, as they tend to create voltage instability by preventing excess electrode voltage that arises during dynamic clamp oscillation from decaying before the next time step occurs. Increasing dynamic clamp time step performance increased this stability, but only when electrode compensation was sufficient. Unfortunately, the methods for attaining the optimal level of electrode compensation have yet to be developed, so the presence of this instability at high conductance outputs should be managed according to the limitations of the system as best as possible in the meanwhile.

5.3 Experimental Applications of Dynamic Clamp

The examples to follow will illustrate some of the basic conceptual questions that the dynamic clamp can be employed to address. The first will illustrate how dynamic clamp can be used to explore the contribution of specific neuronal mechanisms and conditions (such as those contributed by *in vivo*-like conditions or specific type of plasticity in synaptic input) to overall neuronal activity through knockin or knockout of artificial conductances. The second will demonstrate the dynamic clamp's capability to construct *ad hoc* neuronal circuits to investigate the behavior of neuronal networks. These are but a few examples of dynamic clamp's applicability among many in the fields of neurophysiology and cardiac electrophysiology. In addition to these abilities, and the previously mentioned role in creating artificial gap junctions, dynamic clamp has also been used to simulate the effects of other voltage-independent ion channels, voltage-dependent ion channels, and the effects of neuromodulators [7, 13, 23, 24]. We will not cover all the possible applications here, but a number of comprehensive overviews are available [1, 10, 25, 26].

Notice that in all of these experimental possibilities, artificial adjustment of functional membrane conductance serves to empower the experimenter to quickly mimic physiological conditions that would otherwise be possible only with more time-consuming and expensive pharmacological and genetic techniques, and does so with far less limitations on what is possible in the modern laboratory. This is due to the general applicability of the conductance modeling process. However, also note that there are limitations that arise from the very process that enables this generality, as alluded to earlier.

5.3.1
Example Application 1: Neuronal Gain Control

5.3.1.1 Synaptic Background Noise Mechanism

Neurons are known to perform important computational tasks. Two primary examples of such observed phenomenon are additive and multiplicative operations on their synaptic inputs. Functionally, these operations correspond to a transform on the input–output firing rate relationship as a horizontal shift or a slope gain (Figure 5.3), where the independent variable is synaptic firing rate or driving current level, and the dependent variable is output firing rate. The output activity of neurons

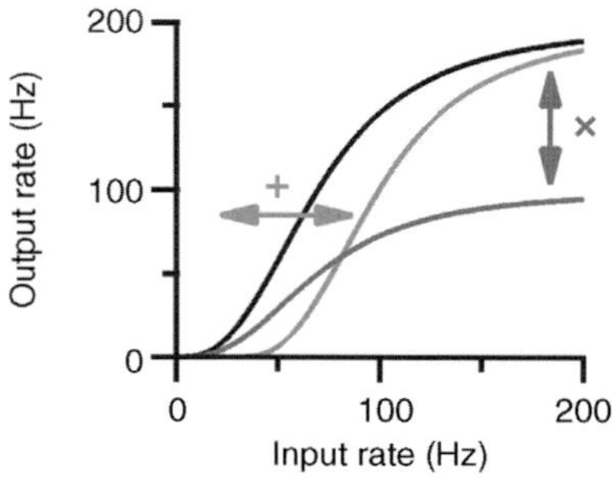

Figure 5.3 Gain modulation in the input–output firing rate relationship of a neuron. A change in gain (dark gray, X) reflects a change in a neuron's input sensitivity, while an additive shift (light gray, +) is typically due to a change in the sum of synaptic inhibition and excitation. Reproduced from Rothman, Cathala, Steuber, and Silver (2009) [36] with permission from Nature Publishing Group.

is often thought of as arising from the sum of synaptic inputs, both excitatory and inhibitory, resulting in enhancement or suppression of output activity based on the overall balance of opposing independent synaptic contributions (often described as the "push–pull" tally). But gain modulation scales the neuron's sensitivity in a distinctly different manner that is essential for many computations. Some processes that have been shown to depend on gain modulation include spatial perception [27], attention [28, 29], auditory processing [30], and coordinate transformations [31–34]. However, lacking a thorough understanding of the mechanism(s) that give rise to this phenomenon, it remained unclear if the traditional view of the push–pull mechanism as the primary mode of information flow to neurons truly captures the full picture of computational throughput in neurons. And it was the dynamic clamp that was uniquely suited to explore this question.

Chance *et al.* [35] used dynamic clamp to examine cortical layer 5 pyramidal neurons and found one such mechanism. In this study, the authors examined the changes in input–output relationship due to varying levels of synaptic background noise. It was known that the presence of this synaptic background activity caused an increase in overall conductance and discharge variability, but the authors discovered that as the strength of this background activity is varied, gain is modulated. This was discovered by introducing model synaptic activity that mimics *in vivo* conductance fluctuations into *in vitro* single neurons with dynamic clamp. Because the imposed conductance waveform is modeled on a computer, its parameters are under full experimental control.

The conductance waveform to be used by the dynamic clamp was constructed by first determining the appropriate waveform for the unitary excitatory and inhibitory conductance that arises from each presynaptic activation of AMPA or $GABA_A$ ion channel populations, then summing these presynaptic events timed on a spike train generated with a Poisson process. This was repeated independently for excitatory and inhibitory channels, and finally injected into the cell after the dynamic clamp converted the conductance waveforms into the appropriate current by multiplying the conductance waveforms by their respective synaptic driving force, taken as the difference of the synaptic reversal and membrane potentials,

$$I_{syn} = g_E(E_E - V_m) + g_I(E_I - V_m).$$

Here, g_E and g_I are the excitatory and inhibitory conductance waveforms, respectively, E_E and E_I are the excitatory and inhibitory synaptic reversal potentials, respectively, and V_m is the recorded membrane potential.

The authors balanced the excitatory and inhibitory synaptic currents so that the observed changes in input–output relationship with changes in background synaptic activity would be free from confounding effects of altered overall excitatory drive that would otherwise occur, resulting in little overall depolarization or hyperpolarization. By varying the background synaptic input firing rates proportionally for both excitatory and inhibitory inputs over discrete steps of $0\times$ (no background synaptic input), $1\times$, $2\times$, and $3\times$, the baseline observed synaptic background firing rate, then driving the cell with a series of constant currents, the input–output relationship was observed and plotted. The results showed consistent incremental change in firing rate sensitivity to driving current with changes in background synaptic rate, with no observed additive shifts in preparations that did not exhibit excess excitation or inhibition. These results correspond well to results from a computational model. If indeed this is a primary means of gain modulation, the simultaneous rise and fall of synaptic input signals' excitatory and inhibitory components would have to be a neural phenomenon that carries a signal to command changes in neural gain sensitivity. If this were the case, the level of this covarying push–pull synaptic input could be considered an independent information pathway to control the computation of a neuron, resulting in a mode of operation that is similar to gate control in transistors.

5.3.1.2 **Synaptic Depression Mechanism**

In a subsequent study, Rothman *et al.* [36] used dynamic clamp to find another mechanism that controls neuronal gain that operates independent of noise. They studied cerebellar granule cells (GCs) with dynamic clamp and found that when excitatory synaptic input with short-term depression (STD) is present, gain modulation is mediated by inhibitory input. In this system, the GC is synaptically excited by approximately four mossy fibers (MFs) and receives inhibitory input from Golgi cells. The authors replaced this synaptic input with model synaptic input via dynamic clamp, which created the ability to experimentally control the synaptic depression and firing rate with model adjustments.

In order to construct the model synaptic excitatory postsynaptic currents (EPSCs) for the dynamic clamp to inject and modulate, EPSCs were evoked by extracellular stimulation of single MF inputs with Poisson stimulation trains, and the resulting GC output was recorded. Trains with similar statistics were repeated such that an average waveform could be calculated, and stimulus artifacts were removed. This formed the model for the depressing EPSC waveform once the voltage recording was converted to conductance. Nondepressing waveforms were then constructed by fitting the first EPSC responses in each train to a multiple exponential function, then convolving this model with the original Poisson stimulation train. The results of introducing these two different artificial conductance trains showed a modulation of gain by STD that occurs independent of noise, and appears to arise from the nonlinearity in time-averaged excitatory synaptic conductances (G_{exc}) as a function of MF stimulation rate introduced by depression since the GC firing rate does not change at any given G_{exc} in the presence or absence of STD. Furthermore, the magnitude of this change in gain was shown to be mediated by inhibitory input using the dynamic clamp to introduce simple tonic inhibition. Each of these phenomena was brought to light with dynamic clamp and reproduced with computational model simulations.

5.3.2
Example Application 2: Constructing Artificial Neuronal Circuits

We have already seen how one of the dynamic clamp's most prominent features is its ability to create artificial synapses in electrically excitable cells. Because the manifestation of this synapse consists of a mathematical expression executed by a hardware platform interfacing with either a computational model cell or real cell, and all of the parameters and connections are specified by the experimenter, this technique can be used to establish any circuit that the user desires, within the constraints of the hardware platform and experimental preparation accessibility limitations (see Section 5.4). Just as any pair of electrically excitable cells can be coupled by independently interfacing each individual cell with the same dynamic clamp and accounting for the recorded activity of each cell in the equation of the model synapse, any larger set of cells can be connected to one another by extending this procedure to multiple cells. This not only allows the construction of biological and hybrid circuits that would not otherwise have existed, it also gives the user control over the structure and parameters of the circuit created, and an additional avenue for inspection of the operating principles of circuits of electrically excitable cells.

One example of such a study focused on exploring the intercircuit synchronization of microcircuits [37], such as those that are proposed to underlie the complex cortical activity in the brain. In one microcircuit view of brain activity, cortical microcircuits exist with stereotyped topology that is composed of a group of neurons that interact with one another through synapses and result in an overall circuit activity like that of a central pattern generator (CPG), which is characterized by oscillatory behavior. These microcircuits in turn are connected by synaptic connections in a manner that gives rise to intercircuit coordination; however, the connections that give rise to this are difficult to assay directly with experiments in the brain. To explore these connections and attempt to determine the most likely connectivity patterns that these intercircuit synapses must display in order to reliably yet flexibly facilitate coordination, the authors connected two invertebrate stomatogastric (STG) CPGs with varied dynamic clamp synapses.

The STG CPG has a circuit configuration similar to that of a cerebral microcircuit, including a pacemaking core connected to inhibitory neurons with reciprocal connections. Synapse configurations were established, one at a time, as either electrical coupling of pacemakers or inhibitory neurons, or chemical synapses simulating excitatory or inhibitory synapses, connecting in unilateral or bilateral configurations. The bilateral chemical synapses were arranged with either direct reciprocal connections between similar neuron types or with indirect contralateral connections from inhibitory neurons to pacemakers (Figure 5.4).

The resulting activity from these experiments indicated that the contralateral inhibitory configuration is the most stable and flexible, offering both an improvement in activity variability over isolated CPGs and an optimality in the coordination of phase relationships between the activity of different CPGs. Electrically coupled and reciprocally inhibiting pacemakers also displayed coordinating ability, but with less stability and flexibility.

In connecting multiple independent neuronal networks to one another, dynamic clamp allowed the experimentalists to explore more general network operating principles. Perhaps in the future, a more complete approximation of the dynamics of cortical microcircuits, which contain a larger population of neurons than a single invertebrate CPG, will be possible with a similar *in vitro* preparation that more

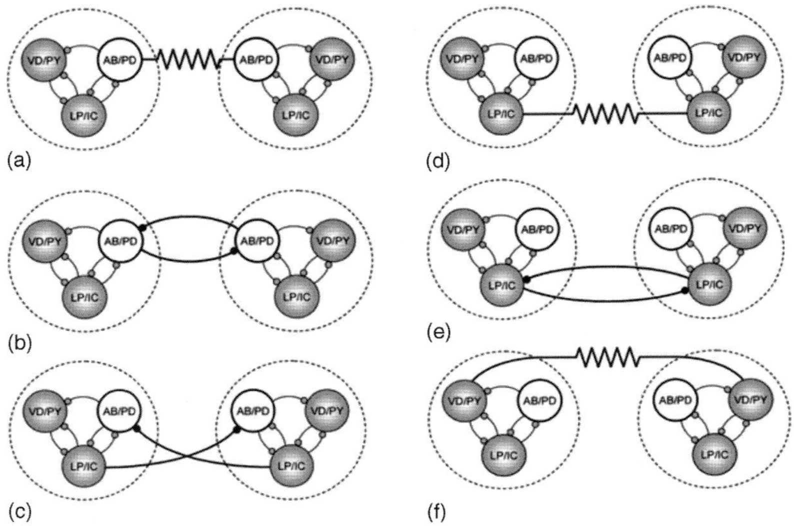

Figure 5.4 Artificial microcircuit configurations constructed from two simplified invertebrate stomatogastric ganglion (STG) pacemakers using dynamic clamp to test microcircuit synchronization. Constituents are the anterior burster (AB) and pyloric dilator (PD) complex, pyloric (PY) and ventricular dilator (VD) neurons, and laterlal pyloric (LP) and inferior cardiac (IC) neurons. Configuration (a) uses electrically coupled pacemakers, and exhibits some activity stability. (b) employs reciprocal inhibitory synapses between pacemakers, while (c) uses contralaterally inhibitory connections between pacemakers and inhibitory neurons, making for the most stable tested configuration. Configurations (d), (e), and (f) displayed the least efficient synchronization. Reproduced from Szücs, Huerta, Rabinovich, and Selverston (2009) [37] with permission from Elsevier.

fully enlists the computational capabilities of the dynamic clamp by introducing computational model neurons into the network, creating a thoroughly hybrid neuronal network.

5.4 Dynamic Clamp System Implementations and Future

5.4.1 Fundamental Considerations

As we have seen, a dynamic clamp implementation may consist of any electronic system that is capable of making a few simple recordings, current commands, and computations. However, the feature set designed into the system determines how rudimentary or versatile it will be. Cost is another issue. And as we will see, performance is a vital consideration. The number and degrees of freedom of these variables has resulted in a large number of different implementations [25], from the original implementations created with analog circuits [2] to the latest systems that run on high-performance personal computers and digital devices.

As personal computer platforms continue to progress and expand in computational power and decrease in cost, laboratories are increasingly relying on these flexible systems to implement dynamic clamp. These systems make a lot of sense for many labs because they are ubiquitous, cost-effective, fast, portable, intuitive to interface with, possess extensive graphical user interfacing (GUI) capabilities and flexibility, and provide the ability to draw on substantial existing technological and user-community resources. Freely available open source software in the real-time operating system and data acquisition card driver suite projects such as RTAI and COMEDI have been particularly useful here (see Section 5.5), as have the commercially developed software suites developed by National Instruments, especially the real-time version of LabVIEW [38]. Real-time operating systems have proven to be valuable tools for fully utilizing the capabilities of the personal computer to meet the performance demands of dynamic clamp. They accomplish this by endowing the computer with the ability to more effectively meet certain performance guarantees in performing a computational cycle, reducing jitter to a level that is not possible with standard operating systems due to unpredictable software interruptions and task scheduling. Implementing dynamic clamp on one of these real-time operating systems circumvents these issues by forcing the running dynamic clamp task to always have top execution priority. However, it appears that the problems that necessitate this solution are becoming less of an issue in some systems with recent multicore processor computers, which tend to split the routine operating system tasks and high priority dynamic clamp tasks across the different processor cores [9].

To create such a dynamic clamp setup, an investigator needs little more than a computer with moderate performance by modern standards, a DAQ board and associated connectors chosen to meet expected performance requirements, a simple electrophysiology rig, and the ability to download software. Steady computer and DAQ board performance increases have been raising the ceiling of possible dynamic clamp step rates up to around 50 kHz for an average multicore system with a low-cost DAQ board, and even higher possibilities for latest generation DAQ cards such as the National Instruments M-series cards.

For various reasons, many implementations of the dynamic clamp based on various technologies exist, with varied user interfaces, and different performance advantages and drawbacks. Often individual laboratories would have unique needs that were not readily met by available systems, so developed their own. Some have attempted to make their systems broadly applicable and available for the research community. Unfortunately, we cannot discuss all implementations in detail here, but as this topic has been covered previously [10, 25], we will discuss a few of the implementations released since the last overview. We also have attempted to list all the available systems in the chapter resources.

5.4.2
Recent and Future Implementations

Of the many personal computer implementations that exist, some recent efforts to push the boundaries of what these systems are capable of, as well as increase their ease of installation and use, have been solidifying. The Real-Time eXperiment Interface (RTXI) software package is now an active, well supported, in-use dynamic clamp (see resources). This project has been the result of a combined effort of three labs that had previously independently developed real-time Linux computer dynamic clamp implementations (RTLab [39], RTLDC [40], and MRCI [41]), and has resulted in a much improved, more broadly applicable, and up-to-date system. One of the major drawbacks of real-time Linux-based dynamic clamp systems, which have otherwise been attractive because of their low cost, precision, stability, and adaptability, has been their sometimes significant installation difficulty. This issue is being addressed in RTXI through the development of a live CD that also works as an installation disk, which is at present functional and available for beta testing. With this development, installation difficulty should no longer be a significant barrier that labs need be concerned about when deciding between different systems. RTXI also adds some new features and flexibility such as a customizable graphical user interface, community-supported plug-in functionality, and libraries that increase ease of programming and flexibility of experimental protocols (see repository on Web site), and file recording in the HDF5 file format (see Section 5.5), which makes saving, managing, and organizing experimental data significantly easier (personal communication, David Christini). However, like most other computer-based dynamic clamp implementations, RTXI still relies on DAQ board technology that limits the number of physical output channels and output speed, and is limited in the model complexity it can handle due to computational limits of contemporary personal computers [42].

Recently, another intriguing computer-based dynamic clamp implementation was developed as an extension of the freely available Microsoft Windows-based QuB software package for electrophysiology (see Section 5.5). Although it does not technically use a hard real-time system, latency and jitter performance have proven to be excellent on multicore systems (personal communication, Lorin Milescu) since dynamic clamp tasks tend to run exclusively on one or more of the available CPUs in a multicore computer with this system. Furthermore, by building on top of a more broadly targeted electrophysiology software suite, this dynamic clamp inherited an impressive library of functionality and GUI flexibility that nicely compliments traditional dynamic clamp operations in a way that makes novel application of dynamic clamp procedures straightforward. For instance, one of this program's first uses was to study kinetic parameters of ion channels in a novel way that combined dynamic clamp simulation of Markov model based ion channels with a parameter search [9].

Still another new dynamic clamp developed by Hugh Robinson at the University of Cambridge is continuing on the path of implementing dynamic clamp on digital signal processing (DSP) hardware to increase performance [25, 43]. This particular development has opened up the possibility of having both the performance benefits of executing dynamic clamp model equations on hardware that is more ideally suited to fast, highly parallel, low-jitter computation than the more generalized CPU processor, while retaining all the benefits of an ease of programming that a personal computer-based dynamic clamp would provide. But directly programming a DSP is typically a time-consuming process that requires some technical expertise, which makes it a less than ideal platform for varied and repeated dynamic clamp experiments. Robinson circumvented this limitation by developing a scripting language that performs the major programming tasks for a given typical dynamic clamp operation, greatly increasing the user-friendliness of such systems for use in dynamic clamp. Further efforts along these lines should one day make such systems as easy to use and as much versatile as the computer-based dynamic clamp systems, thus making accessible their hardware's full potential for significant increases in performance and model and experiment complexity to many. For the moment, the financial expense to acquire such a system remains higher than a computer-based system. This will continue to be an important factor as laboratories evaluate their dynamic clamp needs.

Perhaps among the most exciting current directions in new dynamic clamp implementations are the moves to make the mechanisms of action more biological in nature and broad in experimental applicability. Some of the systems such as RTBiomanager [44] and RTXI are becoming so versatile that they are not only capable of mimicking cellular electrical conductance but can also simulate chemical and mechanical stimuli and other outputs, as well as receive such signals for inputs. This was demonstrated recently when dynamic clamp was employed to mediate mechanosensory stimulation by controlling stepper motors that manipulated the physical orientation of the statocysts of the mullosk *Clione limacina* [45]. The question arises just how far this idea of using real-time control in biological scientific experimentation can be pushed, which implies that the future of the dynamic clamp may be bigger than just the dynamic clamp, itself.

5.5 Resources

- Scholarpedia: Dynamic clamp
  http://www.scholarpedia.org/article/Dynamic_clamp
- Dynamic clamp implementations
  -Advanced Dynamic Clamp [46]/RTBiomanager [44] (Linux-based)
  http://arantxa.ii.uam.es/~gnb/adclamp/

  -DynaClamp [11] (Linux-based)
  http://www.amc.nl/index.cfm?pid=4922

  -Dynamic Clamp [47] (Windows-based)
  http://stg.rutgers.edu/software/

  -DynClamp2/4 [48] (Windows-based)
  http://inls.ucsd.edu/~rpinto/dynclamp.html

-G-clamp v2 [38] (Windows/NI LabVIEW-based)
http://hornlab.neurobio.pitt.edu/

-QuB Dynamic Clamp [9] (Windows-Based)
http://www.qub.buffalo.edu/wiki/index.php/Dynamic_Clamp

-RTXI – Real-Time eXperiment Interface (Linux-based)
http://www.rtxi.org/

-SM-2 Digital Conductance Injection System [8] (Windows/DSP-based)
http://www.cambridgeconductance.com/

-Spike Timing Dependent Plasticity Clamp [49] (DynClamp2-based)
http://inls.ucsd.edu/~nowotny/dynclamp.html

- RTAI – the RealTime Application Interface for Linux from DIAPM
  https://www.rtai.org/
- COMEDI – Linux Control and Measurement Device Interface
  http://www.comedi.org/
- HDF5 – The HDF5 Group
  http://www.hdfgroup.org/HDF5/

References

1 Goaillard, J.M. and Marder, E. (2006) Dynamic clamp analyses of cardiac, endocrine, and neural function. *Physiol. (Bethesda)*, **21**, 197–207.

2 Scott, S. (1979) Stimulation simulations of young yet cultured beating hearts. Dissertation, State University of New York at Buffalo.

3 Tan, R.C. and Joyner, R.W. (1990) Electrotonic influences on action-potentials from isolated ventricular cells. *Circ. Res.*, **67** (5), 1071–1081.

4 Tan, R.C., Osaka, T., and Joyner, R.W. (1991) Experimental-model of effects on normal tissues of injury current from ischemic region. *Circ. Res.*, **69** (4), 965–974.

5 Joyner, R.W., Sugiura, H., and Tan, R.C. (1991) Unidirectional block between isolated rabbit ventricular cells coupled by a variable resistance. *Biophys. J.*, **60** (5), 1038–1045.

6 Hodgkin, A.L. and Huxley, A.F. (1952) A quantitative description of membrane current and its application to conduction and excitation in nerve. *J. Physiol. London*, **117** (4), 500–544.

7 Sharp, A.A., Oneil, M.B., Abbott, L.F., and Marder, E. (1993) The dynamic clamp – artificial conductances in biological neurons. *Trends Neurosci.*, **16** (10), 389–394.

8 Robinson, H.P.C. (2008) A scriptable DSP-based system for dynamic conductance injection. *J. Neurosci. Methods*, **169** (2), 271–281.

9 Milescu, L.S., Yamanishi, T., Ptak, K., Mogri, M.Z., and Smith, J.C. (2008) Real-time kinetic modeling of voltage-gated ion channels using dynamic clamp. *Biophys. J.*, **95** (1), 66–87.

10 Wilders, R. (2006) Dynamic clamp: a powerful tool in cardiac electrophysiology. *J. Physiol. London*, **576** (2), 349–359.

11 Berecki, G., Zegers, J.G., Verkerk, A.O., Bhuiyan, Z.A., de Jonge, B., Veldkamp, M.W., Wilders, R., and van Ginneken, A.C.G. (2005) HERG channel (dys)function revealed by

dynamic action potential clamp technique. *Biophys. J.*, **88** (1), 566–578.
12 Berecki, G., Zegers, J.G., Bhuiyan, Z.A., Verkerk, A.O., Wilders, R., and van Ginneken, A.C. (2006) Long-QT syndrome-related sodium channel mutations probed by the dynamic action potential clamp technique. *J. Physiol. London*, **570**Pt (2), 237–250.
13 Sharp, A.A., Oneil, M.B., Abbott, L.F., and Marder, E. (1993) Dynamic clamp – computer-generated conductances in real neurons. *J. Neurophysiol.*, **69** (3), 992–995.
14 Robinson, H.P.C. and Kawai, N. (1993) Injection of digitally synthesized synaptic conductance transients to measure the integrative properties of neurons. *J. Neurosci. Methods*, **49** (3), 157–165.
15 Hutcheon, B., Miura, R.M., and Puil, E. (1996) Models of subthreshold membrane resonance in neocortical neurons. *J. Neurophysiol.*, **76** (2), 698–714.
16 Dayan, P. and Abbott, L.F. (2001) *Theoretical Neuroscience: Computational and Mathematical Modeling of Neural Systems*, The MIT Press, Cambridge.
17 Mascagni, M.V. and Sherman, A.S. (1998) *Methods in Neuronal Modeling: From Ions to Networks*, 2nd edn (eds C. Koch and I. Segev), The MIT Press, Cambridge.
18 Putzier F I., Kullmann, P.H.M., Horn, J.P., and Levitan, E.S. (2009) Dynamic clamp shows that L-type Ca2 + channel gating, not ion selectivity, drives pacemaking of nigral dopamine neurons. Society for Neuroscience, October 17–21, 2009, Chicago, IL.
19 Sherman-Gold, R. (ed.) (1993) *The Axon Guide for Electrophysiology & Biophysics Laboratory Techniques*, Axon Instruments, Inc., Foster City.
20 Reinhardt, V.S. (2005) A Review of Time Jitter and Digital Systems. Proceedings of the IEEE International Frequency Control Symposium and Exposition, August 29–31, 2005, Vancouver, BC, Canada.
21 Bettencourt, J.C., Lillis, K.P., Stupin, L.R., and White, J.A. (2008) Effects of imperfect dynamic clamp: computational and experimental results. *J. Neurosci. Methods*, **169** (2), 282–289.
22 Preyer, A.J. and Butera, R.J. (2009) Causes of transient instabilities in the dynamic clamp. *IEEE Trans. Neural. Syst. Rehabil. Eng.*, **17** (2), 190–198.
23 Swensen, A.M. and Marder, E. (2001) Modulators with convergent cellular actions elicit distinct circuit outputs. *J. Neurosci.*, **21** (11), 4050–4058.
24 Beenhakker, M.P., DeLong, N.D., Saideman, S.R., Nadim, F., and Nusbaum, M.P. (2005) Proprioceptor regulation of motor circuit activity by presynaptic inhibition of a modulatory projection neuron. *J. Neurosci.*, **25** (38), 8794–8806.
25 Prinz, A.A., Abbott, L.F., and Marder, E. (2004) The dynamic clamp comes of age. *Trends Neurosci.*, **27** (4), 218–224.
26 Destexhe, A. and Bal, T. (eds) (2009) *Dynamic-Clamp: From Principles to Applications*, Springer, New York.
27 Andersen, R.A., Essick, G.K., and Siegel, R.M. (1985) Encoding of spatial location by posterior parietal neurons. *Science*, **230** (4724), 456–458.
28 McAdams, C.J. and Maunsell, J.H.R. (1999) Effects of attention on orientation-tuning functions of single neurons in macaque cortical area V4. *J. Neurosci.*, **19** (1), 431–441.
29 Treue, S. and Trujillo, J.C.M. (1999) Feature-based attention influences motion processing gain in macaque visual cortex. *Nature*, **399** (6736), 575–579.
30 Ingham, N.J. and McAlpine, D. (2005) GABAergic inhibition controls neural gain in inferior colliculus neurons sensitive to interaural time differences. *J. Neurosci.*, **25**, 6187–6198.
31 Zipser, D. and Andersen, R.A. (1988) A back-propagation programmed network that simulates response properties of a subset of posterior parietal neurons. *Nature*, **331** (6158), 679–684.
32 Brotchie, P.R., Andersen, R.A., Snyder, L.H., and Goodman, S.J. (1995) Head position signals used by parietal neurons to encode locations of visual-stimuli. *Nature*, **375** (6528), 232–235.
33 Salinas, E. and Abbott, L.F. (1995) Transfer of coded information from sensory to motory networks. *J. Neurosci.*, **15** (10), 6461–6474.

34 Pouget, A. and Sejnowski, T.J. (1997) Spatial transformations in the parietal cortex using basis functions. *J. Cogn. Neurosci.*, **9** (2), 222–237.

35 Chance, F.S., Abbott, L.F., and Reyes, A.D. (2002) Gain modulation from background synaptic input. *Neuron*, **35** (4), 773–782.

36 Rothman, J.S., Cathala, L., Steuber, V., and Silver, R.A. (2009) Synaptic depression enables neuronal gain control. *Nature*, **457** (7232), 1015–1018.

37 Szucs, A., Huerta, R., Rabinovich, M.I., and Selverston, A.I. (2009) Robust microcircuit synchronization by inhibitory connections. *Neuron*, **61** (3), 439–453.

38 Kullmann, P.H.M., Wheeler, D.W., Beacom, J., and Horn, J.P. (2004) Implementation of a fast 16-bit dynamic clamp using LabVIEW-RT. *J. Neurophysiol.*, **91** (1), 542–554.

39 Christini, D.J., Stein, K.M., Markowitz, S.M., and Lerman, B.B. (1999) A practical real-time computing system for biomedical experiment interface. *Ann. Biomed. Eng.*, **27** (2), 180–186.

40 Dorval, A.D., Christini, D.J., and White, J.A. (2001) Real-time linux dynamic clamp: a fast and flexible way to construct virtual ion channels in living cells. *Ann. Biomed. Eng.*, **29** (10), 897–907.

41 Raikov, I., Preyer, A., and Butera, R.J. (2004) MRCI: a flexible real-time dynamic clamp system for electrophysiology experiments. *J. Neurosci. Methods*, **132** (2), 109–123.

42 Hughes, S.W., Lorincz, M., Cope, D.W., and Crunelli, V. (2008) NeuReal: an interactive simulation system for implementing artificial dendrites and large hybrid networks. *J. Neurosci. Methods*, **169** (2), 290–301.

43 Lien, C.C. and Jonas, P. (2003) Kv3 potassium conductance is necessary and kinetically optimized for high-frequency action potential generation in hippocampal interneurons. *J. Neurosci.*, **23** (6), 2058–2068.

44 Muñiz, C., de Borja Rodríguez, F., and Varona, P. (2009) RTBiomanager: a software platform to expand the applications of real-time technology in neuroscience. Eighteenth Annual Computational Neuroscience Meeting: CNS*2009, 18–23 July 2009, Berlin, Germany.

45 Muniz, C., Levi, R., Benkrid, M., Rodriguez, F.B., and Varona, P. (2008) Real-time control of stepper motors for mechano-sensory stimulation. *J. Neurosci. Methods*, **172** (1), 105–111.

46 Muniz, C., Arganda, S., Rodriguez, F.D., and de Polavieja, G.G. (2005) Realistic stimulation through advanced dynamic-clamp protocols. *Lect. Notes Comput. Sci.*, **3561**, 95–105.

47 Rabbah, P. and Nadim, F. (2005) Synaptic dynamics do not determine proper phase of activity in a central pattern generator. *J. Neurosci.*, **25** (49), 11269–11278.

48 Pinto, R.D., Elson, R.C., Szucs, A., Rabinovich, M.I., Selverston, A.I., and Abarbanel, H.D.I. (2001) Extended dynamic clamp: controlling up to four neurons using a single desktop computer and interface. *J. Neurosci. Methods*, **108** (1), 39–48.

49 Nowotny, T., Szucs, A., Pinto, R.D., and Selverston, A.I. (2006) StdpC: a modern dynamic clamp. *J. Neurosci. Methods*, **158** (2), 287–299.

6
Biohybrid Circuits: Nanotransducers Linking Cells and Neural Electrodes

Linfeng Chen, Jining Xie, Hargsoon Yoon, Malathi Srivatsan, Robert E. Harbaugh, and Vijay K. Varadan

6.1
Introduction to Neural–Electrical Interfaces

The nervous system operates with electrical signals that are generated by neurons and conducted through a complex network. Since the 1950s, there has been growing interest in developing electrical interfaces to the nervous system. Such neural–electrical interfaces connect neurons, the electrically active cells of the nervous system, and electronic circuitry, and are used for neural recording and stimulation in neuroscience research to understand the physiological processes at the cellular level, and in neural prosthetics to help restore functions in the nervous system [1]. In these applications, the most common transduction scheme is based on microelectrodes, which work as a bridge between nerve cells and outside equipment for transferring charge.

With rapid advances in the past few decades, microelectrodes are now considered an important component of treatment systems for neurological disorders such as Parkinson's disease [2], epilepsy [3], and Tourette syndrome [4]. Their application is also extended to neural prosthetes, such as retinal implants, cochlear implants, and for urinary bladder function restoration, and limb control in paralyzed individuals [5–10].

6.1.1
Typical Types of Microelectrode Arrays

Though single microelectrodes are still in use, in most cases it is highly desirable to record the electrical activity from and/or stimulate multiple neurons simultaneously so that neuronal network physiology can be studied. Therefore, much effort has been made in developing various types of microelectrode arrays (MEAs) for different purposes. Generally speaking, the MEAs for recording signals from and stimulating neurons fall into two categories: planar MEAs and neural probes. Figure 6.1a shows an example of a planar MEA for neurons in culture. In such a structure, the electrode surface is approximately a flat two-dimensional (2D) surface, and there is a well

Biohybrid Systems: Nerves, Interfaces, and Machines, First Edition. Edited by Ranu Jung.

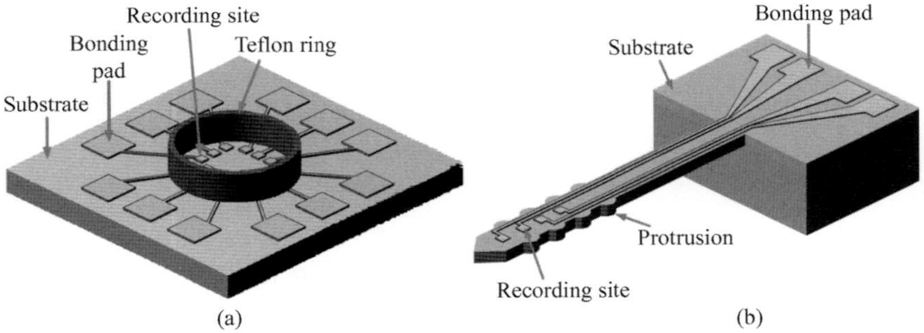

Figure 6.1 Examples of microelectrodes for recording from and stimulating neurons. (a) Planar MEA for *in vitro* recording from neurons in culture. (b) Neural probe for *in vivo* cerebral neural recording. Reproduced with permission from Paik *et al.* [11]. Copyright 2003 IOP Publishing Ltd.

(Teflon ring) for cell culture. Planar MEAs are widely used for *in vitro* studies of neural network physiology. Several companies, such as Alpha MED Scientific Inc. (Japan), Multi Channel Systems MCS GmbH (Germany), Bio-Logic SAS (France), and Ayanda Biosystems SA (Switzerland), can provide planar MEAs for both stimulation and acquisition purposes. Planar MEAs with low impedance (less than 30 kΩ) and different electrode spacing can be obtained from the Alpha MED Scientific Inc. The Multi Channel Systems MCS GmbH can provide MEAs with different electrode spacing and different numbers of channels and wells. The combination of Bio-Logic SAS and Ayanda Biosystems SA can provide MEAs with a three-dimensional (3D) electrode shape that allows better tissue penetration and minimizes the distance to active cells within the tissue.

Figure 6.1b shows an example of a neural probe. In this structure, multiple recording and/or stimulating sites are located on a sharp protrusion, and the probe can be used for *in vivo* recording and stimulation. Three-dimensional arrays of recording and/or stimulating sites can be achieved by arranging protrusions with desired stimulating and/or recording sites. Using such 3D neural probes, detailed mapping of connections in the nervous system could be possible, and such studies will give important insights into the signal processing mechanisms in the nervous system [12]. Most of the current neural probes are silicon based, and they were made using standard planar photolithographic CMOS-compatible techniques on silicon wafers. Michigan array and Utah array are two typical kinds of silicon-based neural probes that are widely used in neuroscience research and medical practice [1]. Neural probes with different specifications are also commercially available. For example, the NeuroNexus Technologies Inc. (USA) can provide neural probes, which are microfabricated thin film MEMS devices with a typical thickness of 15 μm and a typical width of 150 μm. Such a probe has up to 64 channels and can target brain regions up to 10 mm deep; therefore, simultaneous sampling of various cortical brain layers can be made.

Although rigid MEAs have long been used, the mechanical mismatch between the stiff probe and soft biological tissues may aggravate inflammation at the implantation site. In recent years, flexible substrates, such as polyimide, have been used for reducing the chronic tissue inflammation responses [1, 13]. In addition, MEAs on flexible substrates do not easily break during biological and medical reliability testing.

6.1.2
Electric Circuit Model

When microelectrodes are used to record neural signals, they are immersed in a conducting medium, and electrochemical reactions occur at the interface between the solid microelectrodes and the electrolyte. Microelectrodes record the voltages produced by ionic current flow around neurons as their cell membranes depolarize in response to inputs received from other cells. These neural "spike" potentials represent the electrical half of an electrochemical system, with amplitudes as high as several hundred microvolts and frequencies up to about 10 kHz [12, 14].

Figure 6.2 shows an electric circuit model for the electrode–electrolyte interface, including the interfacial capacitance (C_I), the charge transfer resistance (R_t), the diffusion-related Warburg elements (R_W and C_W), and the solution resistance (R_s) [15].

Although the parameters in the model are related to some physical properties of the electrode–electrolyte interface, they can be scaled with frequency and electrode areas. The impedances scaling with the frequency (f), the surface area (A_S), and the geometric area (A_G) are given by [11]

$$|Z_{C_I}| = k/(fA_S), \tag{6.1}$$

$$|Z_{R_t}| = k/(fA_S), \tag{6.2}$$

$$|Z_W| = k/(\sqrt{f}A_S), \tag{6.3}$$

$$|Z_{R_S}| = k/(\sqrt{f}\sqrt{A_G}), \tag{6.4}$$

where k is a constant determined by the electrochemistry and ion mobility. The above equations indicate that an increase in A_S or A_G will decrease the impedance of the electrode–electrolyte interface.

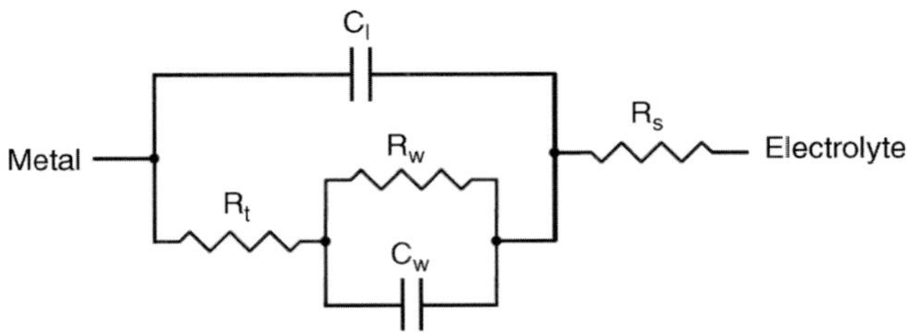

Figure 6.2 Circuit model for an electrode–electrolyte interface.

An electrode–electrolyte interface with low impedance is critical for neural recording and stimulation. For *in vitro* monitoring of electrogenic cells, where small microelectrodes are required for high-resolution recording and stimulation, the need for low interface impedance is twofold. For stimulation, a certain current density is necessary to generate activity. High impedance would result in a large applied electrode voltage leading to undesirable electrochemical reactions that may be harmful to neurons in culture. For recording, as the extracellular signals are low, on the order of microvolts for neurons, the neural signals will be lost in the noisy, ion-based electric fluctuations of the surrounding electrolyte media if the electrode impedance is not low enough [16]. In neural prosthetic systems, low-impedance electrode–tissue interfaces are important both for maintaining signal quality for recording and for effective charge transfer for stimulation [17].

In general, a recording microelectrode has a surface area in the range of 100–400 μm^2, and a stimulating microelectrode with charge injection density around 800 $\mu C/cm^2$ has a surface area in the range of 1000–10 000 μm^2 [12, 18]. Usually, microelectrodes have high impedance due to their small surface areas. Among various methods for reducing interface impedances, based on the considerations of signal-to-noise ratio, process steps, cell size, and positioning of probes, it is highly desirable to minimize the interface impedance by increasing the effective areas of microelectrodes, without increasing the physical surface area. The effective area of an electrode can be increased through micropatterning, electroplating, surface roughening, chemical modification, and so on [11, 19–21].

6.1.3 Requirements on Electrode Materials

In the development of MEAs, to ensure long-term sensitive recording and effective stimulation, there are special requirements on the electrode materials:

1) **Excellent cell adhesion**: The surface roughness plays a critical role in neuron adhesion [22]. Materials with rough surfaces or materials whose roughness could be significantly increased by traditional surface roughening techniques are favorable.
2) **Strong capability of transferring electrical signals**: In an MEA system, transduction is accomplished through the conversion processes between electrophysiological activities of neurons and electrical signals. An MEA electrode must be electrically conductive and it should conduct the desired electrical current to the contacting neurons in milliseconds to avoid the electrolysis of both water and biomolecules [23].
3) **High signal-to-noise ratio and high effectiveness**: As discussed in Section 6.1.2, increasing the effective area of a microelectrode is an effective way for decreasing the interface impedance of the microelectrode. Therefore, materials whose effective surface areas can be greatly increased by physical, chemical, and electrical techniques are desirable.
4) **Long-term operation**: For neural implant applications, an MEA should allow long-term recording and/or stimulation. Unexpected chemical reactions with the biological medium may lead to a complete failure for the MEA operation.

Due to this reason, a combination of good biocompatibility, mechanical stability, and chemical durability is required for suitable MEA material candidates.

Based on the above considerations, inert metals, such as platinum, gold, iridium, and tungsten, have been used in the fabrication of MEAs for a long time. Metallic oxide materials, such as iridium oxide and tantalum pentoxide, have also been used to enhance the electrode efficiency and long-term reliability. Meanwhile, conducting polymers including poly(pyrrole) (PPy) and poly(3, 4-ethylenedioxythiophene) (PEDOT) are also investigated seeking the advantage of excellent biocompatibility and high electrode efficiency.

6.1.4
Applications of Nanotechnology

Nanotechnology has a substantial impact on neuroscience because molecular and signal processing occurs at the micrometer scale of neurons, which have distinct nanometer-scale compartments, such as synapses, axons, and dendrites [19, 20]. Recent advances in nanotechnology offer a unique opportunity to integrate nanostructures on the surfaces of electrodes to improve neural interfaces. The integration of nanostructures on an electrode may significantly increase the effective surface area of the electrode and decrease the impedance of the neural–electrical interface.

The advances of nanotechnology have also made it possible to develop nanoelectrodes, which may exhibit unbeatable spatial and temporal resolution as well as sensitivity [24, 25]. This area is still in its infancy, and at present, the main issues facing the development and applications of nanoelectrodes include the preparation of devices and the understanding of their electrochemical performances.

In this chapter, the applications of nanomaterials for improving the neural–electrical interfaces will be discussed. Among various types of nanomaterials, nanoparticles are first used in decreasing the impedance of neural–electrical interfaces. Platinum black (Pt nanoparticle) coating is a common method for integrating nanostructures onto neural electrodes [26]. Although platinum black coating is a simple process that yields a large area increase, this approach is unreliable for long-term implantations because its flimsy dendritic structures are not mechanically robust and they would quickly dissolve and thereby lose their effectiveness. This chapter will concentrate on the applications of nanowires, carbon nanofibers (CNFs), carbon nanotubes (CNTs), and conducting polymer nanomaterials for the improvement of MEAs. The development of nanoelectrodes will also be discussed, and future research directions in this field will be given at the end of this chapter.

6.2
Neural Probes with Nanowires

Based on the merits of combining one-dimensional (1D) behavior of nanowires with the three-dimensional, high aspect ratio columnar structures, the use of aligned nanowires is one of the most active areas of development [27–30]. The aligned nanowires on MEA electrodes may significantly increase the electrochemically active area of the electrodes without losing spatial resolution of the MEA. In the following sections, we discuss neural probes with metallic nanowires and metal oxide nanowires.

6.2.1
Metallic Nanowires for Neural–Electrical Interfaces

Metallic nanowires, such as Au and Pt, have been used for a wide range of biomedical applications. Gold nanowires have been fabricated on various substrates for neural interfaces [19, 31]. As shown in Figure 6.3a, Yoon *et al.* [31] synthesized Au nanowires on microelectrodes with 50 μm diameter on a flexible polyimide substrate. As shown in Figure 6.3b, significant reduction in impedance was observed, compared to planar Au electrodes with the same dimension. As shown in Figure 6.3c, the biocompatibility of nanowire electrodes was confirmed by culturing neuronally differentiating pheochromocytoma cells (PC 12) on the electrode array, showing differentiation into neurons and preferential attachment on nanowire electrodes. These nanowire ensembles are being tested both for extracellular neural recording and stimulation using the brain and for physiological sensing using cardiac myocytes.

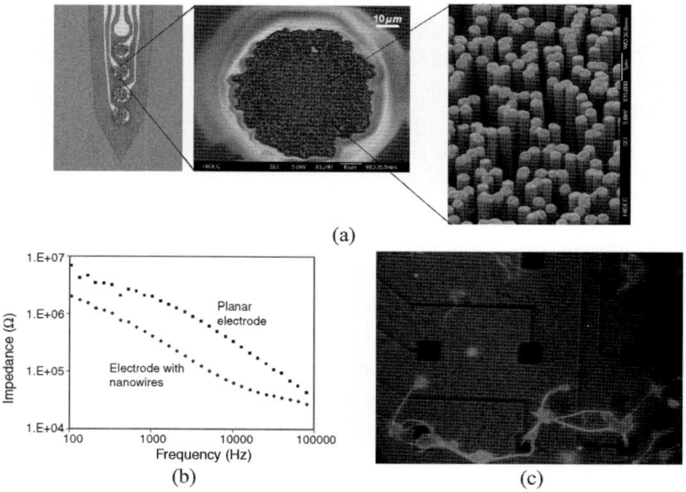

Figure 6.3 (a) SEM image of neural electrode array with Au nanowires on a flexible polymer neural probe. (b) Comparison of impedance magnitude on gold electrodes with planar surface and nanowire bundles measured in PBS buffer at 5 mV AC and 0 V DC potential versus Ag/AgCl reference electrode. (c) Fluorescence microscope image of PC12 cells on nanowire electrode array.

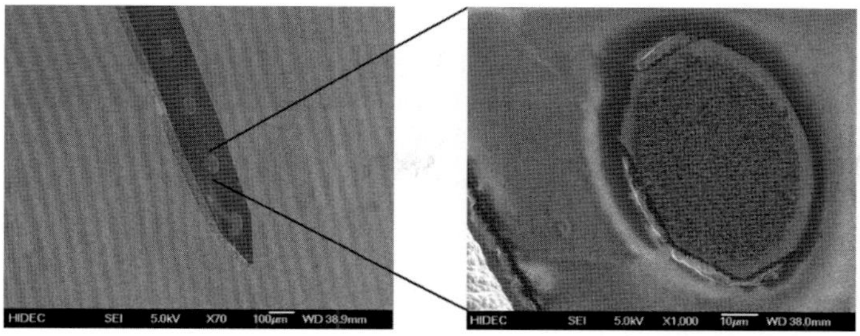

Figure 6.4 Iridium oxide nanowire electrode array on a neural probe.

6.2.2
Metal Oxide Nanowires for Neural–Electrical Interfaces

Several metal oxide thin films have been used to mediate the electrochemical processes at the electrode and cell interface for neural recording and stimulation. Tantalum oxide and iridium oxide have been chosen as electrode materials. Guyton and Hambrecht [9] investigated purely capacitive electrodes with a tantalum pentoxide surface for safe stimulation of nerve cells and tissues without generating electrochemical by-products. The large available surface area and the thin insulating coating of tantalum pentoxide can operate as capacitor-stimulating electrodes having high capacitance per unit volume. However, the application of Ta_2O_5 is limited by low charge injection that results from the planar electrode structure. Instead, an iridium oxide thin film with a pseudocapacitive property shows large charge storage capacity of over 25 mC/cm^2. Electrochemical reaction of iridium oxide involves a reversible reduction and oxidation reaction ($Ir^{4+} + e \rightarrow Ir^{3+}$), so that large charge injection and storage capacity can be obtained. Iridium oxide exhibits both electronic and ionic conductivity.

Iridium oxide thin film can be fabricated by many types of processes including electrochemical anodization of iridium, reactive sputtering, thermal decomposition, and electrochemical deposition. To enhance electrode efficiency, a nanotechnology integration method has been developed using heterostructured iridium oxide/gold nanowires [32]. In this, heterostructured nanowires with an inner core of gold and outer surface layer of iridium oxide provide a large charge storage capacity (48.6 mC/cm^2) for neural signal sensing and stimulation. As shown in Figure 6.4, vertically aligned heterostructured nanowires were fabricated at the tip of a neural probe with polymer or titanium substrates. Impedance measurements indicated that the heterostructured nanowire electrodes exhibited almost three orders of impedance decrease, compared to bare gold electrodes.

6.3
Microelectrode Arrays with Carbon Nanofibers

Carbon nanofibers are cylindrical or conical structures with diameters varying from a few to hundred nanometers and lengths ranging from less than 1 μm to several millimeters. Generally speaking, a carbon nanofiber consists of stacked curved graphite layers that form cones or cups, while the internal structure of carbon nanofibers varies and is comprised of different arrangements of modified graphene sheets [33]. Due to their structural similarity to elongated neurites, CNFs, particularly well-aligned ones, have been used as matrixes to guide axonal growth and nerve repair [34, 35].

The inherent sizes and structures of vertically aligned CNFs (VACNFs) make them attractive for MEA development [34–36]. Conventional planar MEAs have electrodes confined to the substrate plane and cannot penetrate into tissues. They are not ideal for investigating acutely sectioned tissue slices and slice cultures because the injured or dead cells and the reactive glia layers at the surfaces of these samples reduce signal propagation from the active cells to the electrodes. To improve signal quality, it is necessary to develop MEAs with 3D electrodes to penetrate into the tissue under study. MEAs with VACNFs have been developed by several groups [35, 37]. Because the electrodes are carbon-based, these arrays have potential advantages over metal electrodes and could enable a variety of future applications as precise, informative, and biocompatible neural–electrical interfaces.

As shown in Figure 6.5, the VACNF MEA consists of a linear array of 40 individually addressed VACNF electrodes, 10 μm in height, spaced 15 μm apart along a total length of 600 μm. Figure 6.5c indicates that individual VACNF electrodes were cone-like in shape, allowing the electrodes to penetrate into the tissue to improve electrical coupling. Figure 6.5b and c was acquired after the array was used for recording from multiple slice cultures, indicating that the VACNF electrodes are mechanically robust and were not sheared during tissue positioning and removal [35].

As shown in Figure 6.6, to demonstrate the electric recording capability of the array, a hippocampal slice was placed on the array crossing the hilus region with electrode 1 in the DG granule cell layer and electrode 40 in the CA3 pyramidal cell layer. The bicuculline methiodide (BIC)-induced epileptiform activities that propagated throughout the hippocampal slice were detected on multiple channels [35].

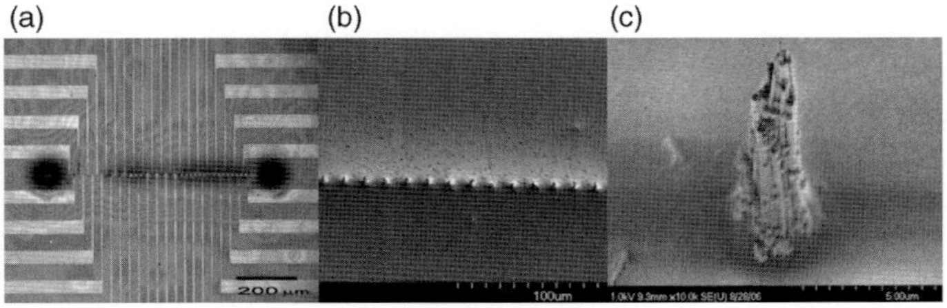

Figure 6.5 Images of VACNF arrays: (a) light micrograph of a VACNF array. (b) SEM image of a part of the VACNF array acquired after several electrophysiological recordings. (c) SEM image of a VACNF electrode from the same array in (b). The SEM images indicate that the VACNF electrodes are mechanically robust and are not sheared by the process of multiple recordings. Reproduced with permission from Yu *et al.* [35]. Copyright 2007 American Chemical Society.

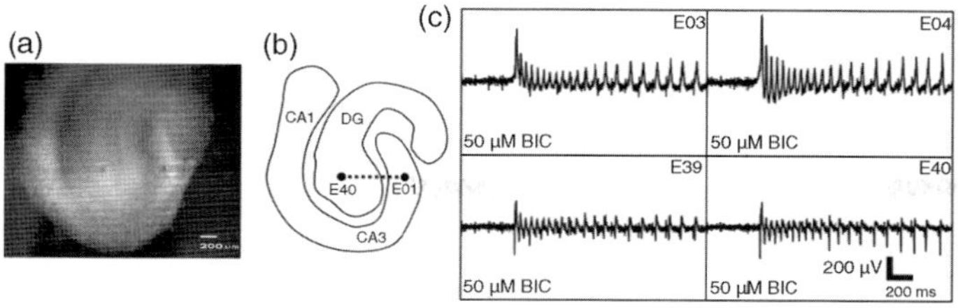

Figure 6.6 Bicuculline-induced epileptiform activity was recorded from hippocampal slices with the VACNF array chips. (a) A light micrograph of a hippocampal slice (13 DIV) on a VACNF array chip. (b) A schematic of the hippocampal anatomy depicts the electrode recording locations. The electrode array crossed the hilus region from the DG granule cell layer to the CA3 pyramidal layer. (c) BIC-induced epileptiform activity as shown for four channels (electrodes 3, 4, 39, and 40). Compared to the spontaneous activity, BIC induced continuous large oscillations with amplitudes up to 600 μV. Reproduced with permission from Yu *et al.* [35]. Copyright 2007 American Chemical Society.

6.4 Microelectrode Arrays with Carbon Nanotubes

A CNT can be imagined as a seamless cylinder rolled up from an sp^2 bonded graphene sheet. Depending upon the number of layers, carbon nanotubes are classified as single-walled carbon nanotubes (SWNTs) and multiwalled carbon nanotubes (MWNTs). The application of carbon nanotubes in neuroscience research started from the first report on the feasibility of using nanotubes as substrates for neuronal growth and as probes of neural functions at the nanoscale [38]. It was found that the surface charge of carbon nanotubes can be used to control the neurite outgrowth [39]. Furthermore, directed neuronal growth can be achieved by using patterned vertically aligned CNTs as the substrate [40].

The feasibility and advantages of using CNTs as MEA electrodes have been confirmed by several groups [41–43]. In the following, we discuss the applications of random CNTs and vertically aligned CNTs in the development of MEAs for neural–electrical interfacing.

6.4.1 Microelectrode Arrays with Random Carbon Nanotubes

A microfabrication process has been used to fabricate MEAs with random CNTs [42, 44]. In this process, TiN tracks were deposited as conductors on a silicon dioxide substrate followed by a deposition of a Si_3N_4 passivation layer. After etching holes on the Si_3N_4 layer, a patterned Ni thin film was deposited as the catalyst for CNT growth. The final process involved a thermal chemical vapor deposition to grow high-density carbon nanotube islands as the MEA electrodes.

When neurons were cultured on such a CNT-based MEA, strong affinity of neurons to CNT islands was observed. As shown in Figure 6.7a, neurons migrated

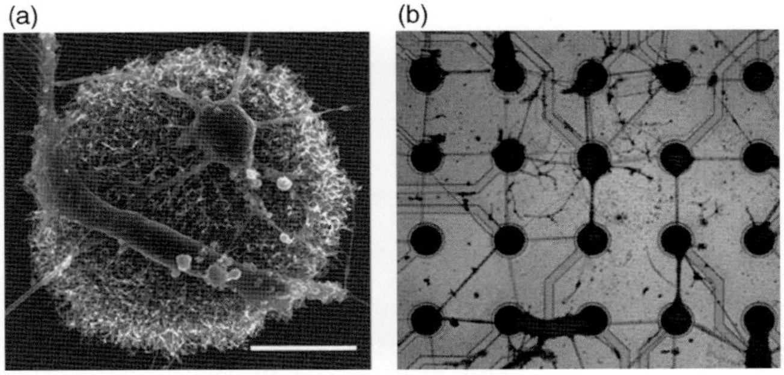

Figure 6.7 (a) SEM micrograph of neuronal cells adhering preferentially to isolated islands of CNTs. (b) Optical microscope image of the patterned interconnected neuronal network formed with CNT-based MEA. Reproduced from Ref. [45] by permission of the Royal Society of Chemistry.

from the flat surface and adhered exclusively onto the CNT electrodes. By culturing neurons with proper density, it was possible to form highly organized neural networks, as shown in Figure 6.7b. This observation suggests a feasible method to engineer the geometry of neural connections in a controlled manner.

A comparative study was conducted on commercial TiN-based MEAs and CNT-based MEAs to investigate their neural recording capabilities [46]. Spontaneous spikes with typical bursting and propagation waves from whole-mount neonatal mouse retinas were recorded consistently after placing retinas over the MEA electrodes. As shown in Figure 6.8, the left traces show the typical unfiltered baseline noise levels while the right traces illustrate spontaneous spiking activities. Obviously, compared to TiN-based electrodes, CNT-based ones display lower baseline noise and substantially higher signals. It was also observed that the spike

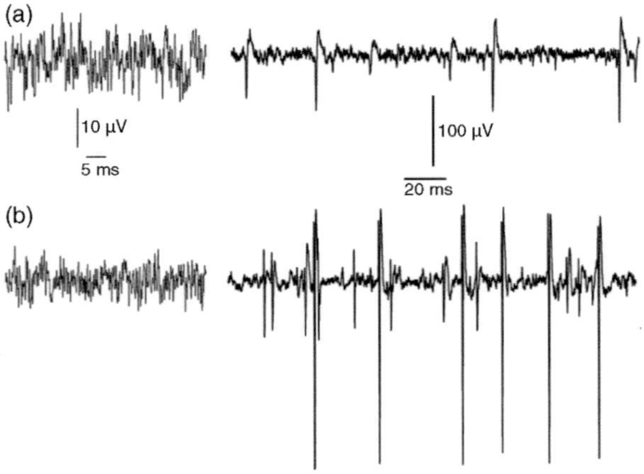

Figure 6.8 Recording neuronal activity by (a) TiN electrodes and (b) CNT-based electrodes. Reproduced with permission from Shoval *et al.* [46].

amplitudes increased gradually over a period, implying improvement of the neuron–electrode coupling.

Gabriel *et al.* [47] reported another type of random CNT-based MEA, which was fabricated by directly depositing SWNT suspension onto standard platinum electrodes. The results showed enhanced electrical properties of the modified MEA electrodes with lower average noises, which can facilitate enormously the identification and isolation of extracellular activities in neural networks.

6.4.2 Microelectrode Arrays with Vertically Aligned Carbon Nanotubes

It appears that random CNT-based MEAs are limited in providing true 3D scaffolds for neuronal growth and cannot offer an effective way for electron transfer. It would be favorable to use vertically aligned CNT arrays for both neural recording and neural stimulation.

Several techniques have been employed for the synthesis of aligned CNTs. In general, it is the van der Waals force among nanotubes that forms the self-oriented large nanotube bundles with sufficient rigidity [48]. A prototype neural interface using vertically aligned CNTs as microelectrodes was developed and proved to be able to enhance the efficacy of neural stimulation and satisfy the safety requirements [43]. The schematic cross section of the device fabricated by microfabrication techniques is depicted in Figure 6.9a. A thin film of heavily doped polysilicon serves as the conducting layer. The top and bottom silicon dioxide and silicon nitride layers are used to provide a hydrophilic surface, to reduce capacitive current, and to block ionic current, respectively. By thermal catalytic chemical vapor deposition, carbon nanotubes self-assemble into uniform pillars projecting orthogonally from the substrate surface, as shown in Figure 6.9b and c. After fabrication, the surface modification of carbon nanotubes is a requisite step for better neuron adhesion. Among various

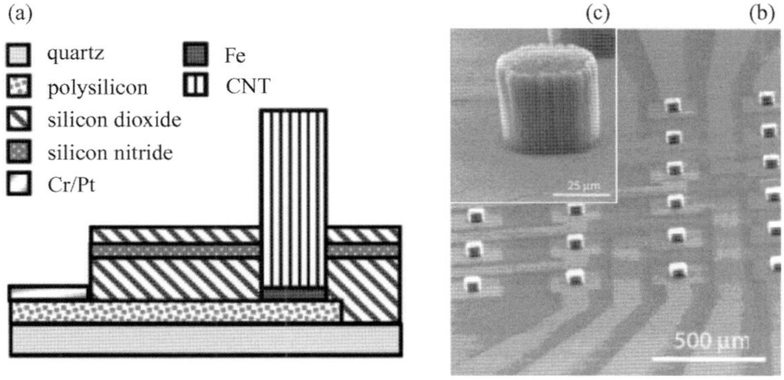

Figure 6.9 (a) Schematic of the vertically aligned CNT pillar microelectrodes. (b) CNT microelectrode arrays. (c) Magnified image. Reproduced with permission from Wang *et al.* [43]. Copyright 2006 American Chemical Society.

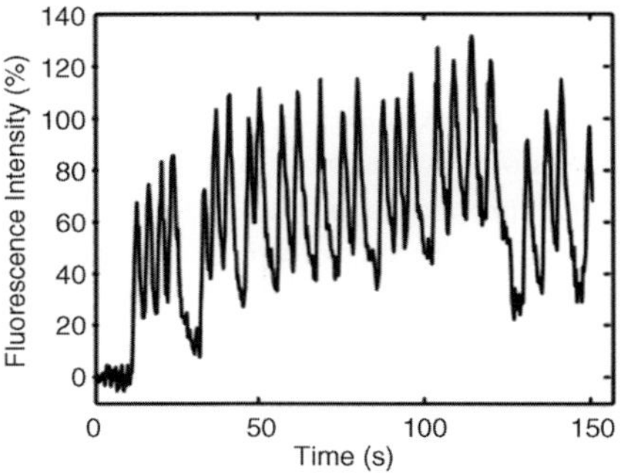

Figure 6.10 Fluorescence intensity change with repeated neuron stimulation. Reproduced with permission from Wang *et al.* [43]. Copyright 2006 American Chemical Society.

methods, noncovalent binding is an appealing approach due to the least invasive characteristics and minimum disturbance to the structures and electronic properties of the nanotubes.

An *in vitro* stimulation was conducted on embryonic rat hippocampal neurons with the vertically aligned CNT-based MEA. After culturing neurons on the MEA for 4 days, current pulses were applied to stimulate neurons through CNT arrays. The action potentials generated by the neurons were detected by observing intracellular Ca^{2+} level change with a fluorescent calcium indicator dye. Thus, by comparing the fluorescence intensity with and without the applied electrical pulses, the efficacy of stimulation can be evaluated. The fluorescent microscope observation confirmed the increase in fluorescence level in hippocampal cells when a stimulus was applied. As shown in Figure 6.10, when a series of monophasic cathodic pulses were applied with intervals, the fluorescence level changed accordingly resulting in multiple peaks. This repeated excitation response strongly suggests both the normal electrode functionality and the good physiological condition of the neuron.

Vertically aligned CNT arrays can be used for neural recording as well. Lin *et al.* [49] developed a flexible aligned CNT-based MEA by microfabrication, and demonstrated the enhanced performance in neural recording. Furthermore, such a flexible device could easily fit the diverse shapes of the human body to record other types of electrical signals such as EMG, EEG, and ECG [49].

6.5 Microelectrode Arrays with Conducting Polymer Nanomaterials

Because of their promising electrical properties, biocompatibility, and potential use for drug delivery, conducting polymers such as PPy, PEDOT, and PEDOT derivatives are attractive for neural interface applications [31, 50, 51]. PEDOT thin films and nanostructures doped with anion such as poly(styrene sulfonate) (PSS) and $LiClO_4$

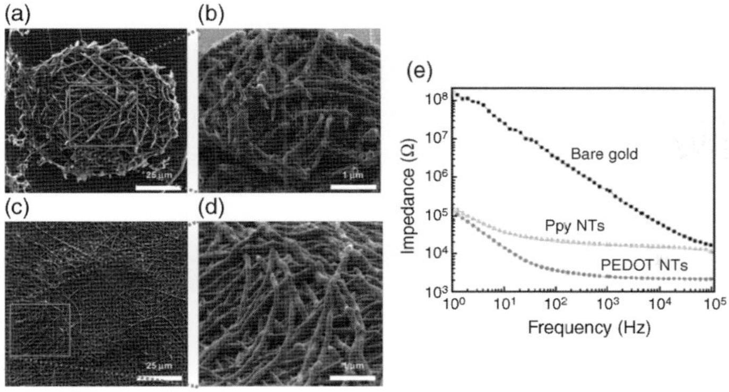

Figure 6.11 Scanning electron micrographs of electropolymerized PPy and PEDOT nanotubes on neural microelectrode sites. (a) Top view of PPy nanotubes, (b) 3D view of PPy nanotubes, (c) top view of PEDOT nanotubes, (d) 3D view of PEDOT nanotubes, and (e) impedance spectroscopy over a frequency range of 1–10^5 Hz. Reproduced with permission from Ref. [52]. Copyright 2008 Elsevier.

have been investigated for the development of MEAs. As shown in Figure 6.11, PEDOT nanotubes were electrochemically polymerized on microelectrodes with electrospun nanofiber template. The impedance measurements showed that the charge storage capacity of the electrode sites modified with PEDOT nanotubes increased by about three orders of magnitude [52].

6.6 Nanoelectrodes for Neural Probes

In neurophysiology, it is generally accepted that sampling of neurons is biased by the tip size of the electrodes used in neural recording, and the signals of many populations of small neurons are rarely recorded with currently available electrodes. Therefore, there could be an advantage of developing nanoscale tipped electrodes. Perhaps, the signals from small neurons that were difficult or almost impossible to be recorded with the standard electrodes would become accessible with smaller tipped electrodes [53].

Furthermore, microelectrodes can only measure extracellular potentials due to their large electrode size. There are other issues associated with extracellular electrodes [54]. First, the large size of electrodes may cause damage to the tissue and cells. Second, signal-to-noise ratio may be too low for meaningful detection. Third, the recorded signal could originate from several cells. These limiting factors might be addressed by intracellular detections using nanoelectrodes. The typical magnitude of extracellular signals is significantly lower than that of intracellular signals (about 100 mV). The significantly higher magnitude of intracellular signals may result in higher measurement accuracy and higher signal-to-noise ratio.

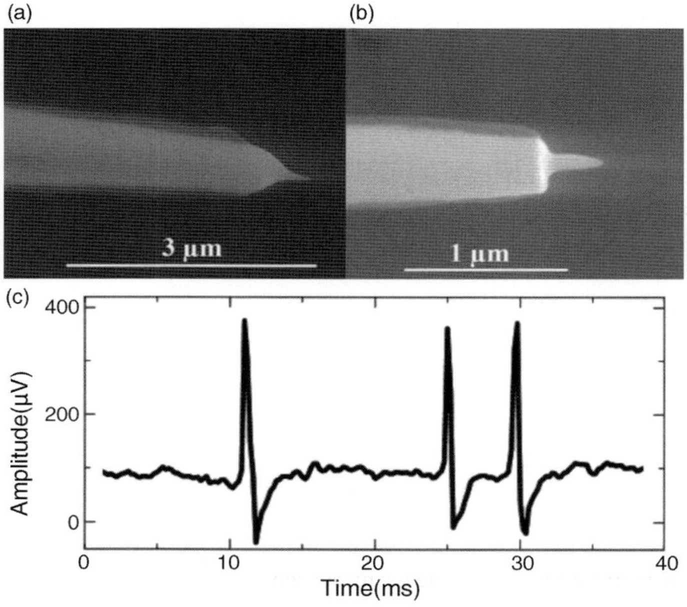

Figure 6.12 (a) Tungsten nanoelectrode with an effective radius of 450 nm. (b) Tungsten nanoelectrode with an effective radius of 100 nm. (c) Trace showing action potentials recorded from an optic tract fiber of a cat with a paint insulated tungsten nanoelectrode (effective radius = 700 nm). Reproduced with permission from Qiao *et al.* [53]. Copyright 2005 IOP Publishing Ltd.

The advancements in nanotechnology have made it possible to fabricate electrodes with diameter less than 1 μm. In the following section, we discuss two types of nanoelectrodes: metal nanoelectrodes and CNT nanoelectrodes.

6.6.1
Metal Nanoelectrodes

Qiao *et al.* [53] developed a procedure for fabricating tungsten nanoelectrodes of consistent tip dimension with high efficiency. The tip of a tungsten wire with diameter 127 μm was etched down to less than 100 nm and then insulated with cathodic electrophoretic paint. Focused ion beam (FIB) polishing was employed to remove the insulation at the apex of the electrode, leaving a nanoscale-sized conductive tip of 100–1000 nm. Figure 6.12 shows SEM images of tungsten nanoelectrodes fabricated by this procedure.

The fabricated electrodes were used to record action potentials extracellularly *in vivo* from single neurons in a cat. For the purpose of demonstration, an electrode with an effective radius of about 700 nm was inserted into the brain of the cat through a guide tube. Electrical signals were recorded with the electrode, amplified by an extracellular amplifier. The nanoelectrode could record well-isolated action potentials from single neurons for more than 50 min, a recording time sufficient for many neurophysiological studies [53]. Figure 6.12c shows the action potentials recorded from an X-cell axon, and the recording exhibits excellent signal-to-noise ratio.

6.6.2 Carbon Nanotube-Based Nanoelectrodes

Based on the mechanical robustness and high conductivity of CNTs, Kawano *et al.* [54] proposed a CNT-based nanoelectrode for intracellular electrophysiological measurements. As illustrated in Figure 6.13a, the nanoelectrode is based on an individual CNT fabricated by a controlled local growth process and subsequently encapsulated with an insulating layer of Parylene C. It is integrated with a silicon microstructure with a total length of 5 μm, and its tip at the distal end is locally heated to expose about 100 nm long CNTas the sensing port. Figure 6.13b shows the TEM images of the tip region. The diameter of the nanotube is 10 nm and the thickness of Parylene C is 50 nm. The inset image is the close-up view of the exposed CNT. Such a CNT nanoelectrode, with its high strength and Young's modulus, may act as a low-invasive intracellular electrode for measurements inside neurons [54].

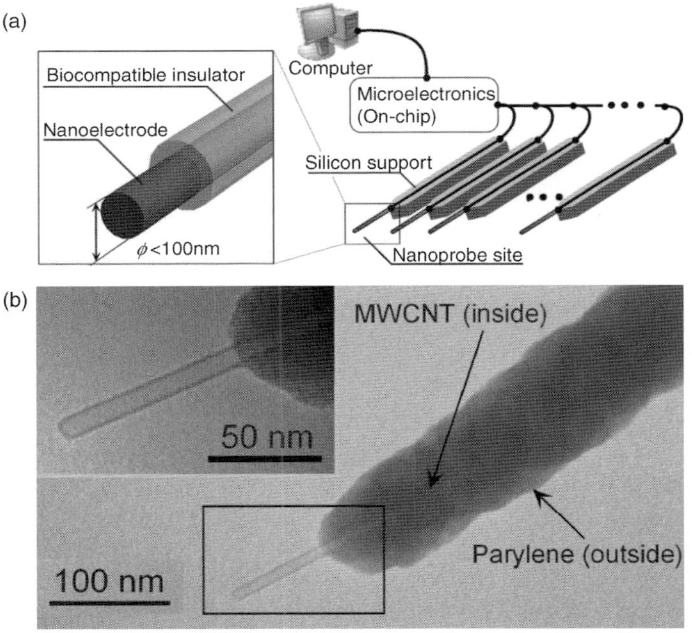

Figure 6.13 (a) Illustration of a carbon nanotube-based nanoprobe with a silicon support. (b) TEM image showing a CNT coated with 50 nm thick Parylene C as an insulator layer. Inset image shows a close-up view of the tip area. The diameter of the CNT is 10 nm. Reproduced with permission from Kawano *et al.* [54]. Copyright 2007 IEEE.

6.7
Summary and Future Work

The extensive efforts in nanotechnology are accelerating the technological evolution in the field of neural–electrical interfaces. Nanotechnology has been used to significantly decrease the impedance of neural–electrical interfaces mainly by increasing the effective surface areas of microelectrodes, resulting in more sensitive recording and more effective stimulation. Nanotechnology has also been used for the development of nanoelectrodes, which could be used to record small neurons, and perform intracellular recording.

Several aspects in the application of nanotechnology for neural–electrical interfaces are attractive for future research in this field. First, there needs to be a complete understanding about neural–nanomaterial interfaces at the nanometer scale. Comprehensive and quantitative characterization of neuron–nanomaterial interfaces is needed, and the robustness, long-term biocompatibility, and functionality of neuron–nanomaterial interfaces should be verified [20]. Second, the chemical aspects of neural recording and stimulation could be addressed through nanotechnology. Recently, extensive attention has been paid to the chemical aspects of the nervous system, with the goal of realizing probes that, in addition to stimulating/recording electrodes, have a complete set of microfluidic components, including chemical sensors, fluidic channels, and valves [12, 21, 51]. Nanomaterials and nanofabrication techniques will play vital roles for this purpose. For example, nanotubes are an ideal candidate for fluidic channels. Third, nanotechnology will be helpful for the development of fully implantable neural microsystems. A fully implantable neural microsystem could simultaneously sample signals at many points in a tissue, providing insight into processes such as movement control, memory formation, and perception [1]. It is expected that nanoelectronics could be integrated into microelectrodes and nanoelectrodes for data processing and signal analysis.

Acknowledgment

This work was supported by the National Science Foundation NSF-EPSCoR Award No. EPS-0701890.

References

1 Cheung, K.C. (2007) Implantable microscale neural interfaces. *Biomed. Microdevices*, **9**, 923–938.

2 Vitek, J.L., Bakay, R.A.E., Hashimoto, T., Kaneoke, Y., Mewes, K., Zhang, J.Y., Rye, D., Starr, P., Baron, M., Turner, R., and DeLong, M.R. (1998) Microelectrode-guided pallidotomy: technical approach and its application in medically intractable Parkinson's disease. *J. Neurosurg.*, **88**, 1027–1043.

3 Cash, S.S., Ulbert, I., Devinsky, O., Bromfield, E., Madsen, J., Cole, A., Halgren, E., and Doyle, W. (2005) Laminar microelectrode recordings of human interictal discharges in neocortical epilepsy reveal complex high-frequency oscillation

patterns. *Epilepsia*, **46** (Suppl. 8), 93–94.

4 Shields, D.C., Cheng, M.L., Flaherty, A.W., Gale, J.T., and Eskandar, E.N. (2008) Microelectrode-guided deep brain stimulation for Tourette syndrome: within-subject comparison of different stimulation sites. *Stereotact. Funct. Neurosurg.*, **86**, 87–91.

5 Horch, K.W. and Dhillon, G.S. (2004) *Neuroprosthetics: Theory and Practice*, World Scientific, New Jersey.

6 Chase, D. (2006) *Shattered Nerves: How Science Is Solving Modern Medicine's Most Perplexing Problem*, The Johns Hopkins University Press, Baltimore, MD.

7 Kim, S.J., Manyam, S.C., Warren, D.J., and Norman, R.A. (2006) Electrophysiological mapping of cat primary auditory cortex with multielectrode arrays. *Ann. Biomed. Eng.*, **34**, 300–309.

8 Loudin, J.D., Simanovskii, D.M., Vijayraghavan, K., Sramek, C.K., Butterwick, A.F., Huie, P., Mclean, G.Y., and Palanker, D.V. (2007) Optoelectronic retinal prosthesis: system design and performance. *J. Neural Eng.*, **4**, S72–84.

9 Guyton, L. and Hambrecht, F.T. (1974) Theory and design of capacitor electrodes for chronic stimulation. *Med. Biol. Eng.*, **12**, 613–619.

10 Lebedev, M.A. and Nicolelis, M.A.L. (2006) Brain–machine interfaces: past, present and future. *Trends Neurosci.*, **29**, 536–546.

11 Paik, S.J., Park, Y., and Cho, D.I. (2003) Roughened polysilicon for low impedance microelectrodes in neural probes. *J. Micromech. Microeng.*, **13**, 373–379.

12 Wise, K.D., Sodagar, A.M., Yao, Y., Gulari, M.N., Perlin, G.E., and Najafi, K. (2008) Microelectrodes, microelectronics, and implantable neural microsystems. *Proc. IEEE*, **96**, 1184–1202.

13 Subbaroyan, J., Martin, D.C., and Kipke, D.R. (2005) A finite-element model of the mechanical effects of implantable microelectrodes in the cerebral cortex. *J. Neural Eng.*, **2**, 103–113.

14 Bard, A.J. and Faulkner, L.R. (1980) *Electrochemical Methods: Fundamentals and Applications*, John Wiley & Sons, Inc.

15 Borkholder, D.A. (1998) *Cell Based Biosensors Using Microelectrodes*, Stanford University.

16 Franks, W., Schenker, I., Schmutz, P., and Hierlemann, A. (2005) Impedance characterization and modeling of electrodes for biomedical applications. *IEEE Trans. Biomed. Eng.*, **52**, 1295–1302.

17 Jimbo, Y. and Kawana, A. (1992) Electrical stimulation and recording from cultured neurons using a planar electrode array. *Bioelectrochem. Bioenerg.*, **29**, 193–204.

18 Cogan, S.F. (2008) Neural stimulation and recording electrodes. *Annu. Rev. Biomed. Eng.*, **10**, 275–309.

19 Zhou, H.B., Li, G., Sun, X.N., Zhu, Z.H., Jin, Q.H., Zhao, J.L., and Ren, Q.S. (2009) Integration of Au nanorods with flexible thin-film microelectrode arrays for improved neural interfaces. *J. Microelectromech. Syst.*, **18**, 88–96.

20 Pancrazio, J.J. (2008) Neural interfaces at the nanoscale. *Nanomedicine*, **3**, 823–830.

21 Cheung, K.C., Djupsund, K., Dan, Y., and Lee, L.P. (2003) Implantable multichannel electrode array based on SOI technology. *J. Microelectromech. Syst.*, **12**, 179–184.

22 Seidlits, S.K., Lee, J.Y. and Schmidt, C.E. (2008) Nanostructured scaffolds for neural applications. *Nanomedicine*, **3**, 183–199.

23 Rose, T.L. and Robblee, L.S. (1990) Electrical stimulation with Pt electrodes. *IEEE Trans. Biomed. Eng.*, **37**, 1118–1120.

24 Arrigan, D.W.M. (2004) Nanoelectrodes, nanoelectrode arrays and their applications. *Analyst*, **129**, 1157–1165.

25 Andrews, R.J. (2007) Neuroprotection at the nanolevel. Part I. Introduction to nanoneurosurgery. *Ann. N. Y. Acad. Sci.*, **1122**, 169–184.

26 Feltham, A.M. and Spiro, M. (1971) Platinized platinum electrodes. *Chem. Rev.*, **71**, 177–193.

27 Delvaux, M. and Demoustier-Champagne, S. (2003) Immobilisation of glucose oxidase within metallic nanotubes arrays for application to enzyme biosensors. *Biosens. Bioelectron.*, **18**, 943–951.

28 Wagner, R.S. and Ellis, W.C. (1964) Vapor–liquid–solid mechanism of single crystal growth. *Appl. Phys. Lett.*, **4**, 89–90.

29 Ren, Z.F., Huang, Z.P., Xu, J.W., Wang, J.H., Bush, P., Siegal, M.P., and Provencio, P.N. (1998) Synthesis of large arrays of well-aligned carbon nanotubes on glass. *Science*, **282**, 1105–1107.

30 Toimil Molares, M.E., Buschmann, V., Dobrev, D., Neumann, R., Scholz, R., Schuchert, I.U., and Vetter, J. (2001) Single-crystalline copper nanowires produced by electrochemical deposition in polymeric ion track membranes. *Adv. Mater.*, **13**, 62–65.

31 Yoon, H., Deshpande, D.C., Ramachandran, V., and Varadan, V.K. (2007) Aligned nanowire growth using lithography-assisted bonding of polycarbonate template for neural probe electrodes. *Nanotechnology*, **19**, 025304.

32 Yoon, H. and Varadan, V.K. (2008) Neural probe developed that will limit damage to cells and biological tissue. Available at http://www.sciencedaily.com/releases/2008/10/081015164334.htm.

33 Melechko, A.V., Merkulov, V.I., McKnight, T.E., Guillorn, M.A., Klein, K.L., Lowndes, D.H., and Simpson, M.L. (2005) Vertically aligned carbon nanofibers and related structures: controlled synthesis and directed assembly. *J. Appl. Phys.*, **97**, 041301.

34 Cassell, A.M., Li, J., Nguyen-Vu, T.B., Koehne, J.E., Chen, H., Andrews, R., and Meyyappan, M. (2009) Vertically aligned carbon nanofibers: interconnecting solid state electronics with biosystems. *J. Nanosci. Nanotechnol.*, **9**, 5038–5046.

35 Yu, Z., McKnight, T.E., Ericson, M.N., Melechko, A.V., Simpson, M.L., and Morrison, B. (2007) Vertically aligned carbon nanofiber arrays record electrophysiological signals from hippocampal slices. *Nano Lett.*, **7**, 2188–2195.

36 Nguyen-Vu, T.D.B., Chen, H., Cassell, A.M., Andrews, R.J., Meyyappan, M., and Li, J. (2007) Vertically aligned carbon nanofiber architecture as a multifunctional 3-D neural electrical interface. *IEEE Trans. Biomed. Eng.*, **54**, 1121–1128.

37 Nguyen-Vu, T.D.B., Chen, H., Cassell, A.M., Andrews, R., Meyyappan, M., and Li, J. (2006) Vertically aligned carbon nanofiber arrays: an advance toward electrical–neural interfaces. *Small*, **2**, 89–94.

38 Mattson, M.P., Haddon, R.C., and Rao, A.M. (2000) Molecular functionalization of carbon nanotubes and use as substrates for neuronal growth. *J. Mol. Neurosci.*, **14**, 175–182.

39 Hu, H., Ni, Y., Montana, V., Haddon, R.C., and Parpura, V. (2004) Chemically functionalized carbon nanotubes as substrates for neuronal growth. *Nano Lett.*, **4**, 507–511.

40 Zhang, X., Prasad, S., Niyogi, S., Morgan, A., Ozkan, M., and Ozkan, C.S. (2005) Guided neurite growth on patterned carbon nanotubes. *Sens. Actuators B*, **106**, 843–850.

41 Lovat, V., Pantarotto, D., Lagostena, L., Cacciari, B., Grandolfo, M., Righi, M., Spalluto, G., Prato, M., and Ballerini, L. (2005) Carbon nanotube substrates boost neuronal electrical signaling. *Nano Lett.*, **5**, 1107–1110.

42 Gabay, T., Jakobs, E., Ben-Jacob, E., and Hanein, Y. (2005) Engineered self-organization of neural networks using carbon nanotube clusters. *Physica A*, **350**, 611–621.

43 Wang, K., Fishman, H.A., Dai, H., and Harris, J.S. (2006) Neural stimulation with a carbon nanotube microelectrode array. *Nano Lett.*, **6**, 2043–2048.

44 Gabay, T., Ben-David, M., Kalifa, I., Sorkin, R., Abrams, Z.R., Ben-Jacob, E., and Hanein, Y. (2007) Electro-chemical and biological properties of carbon nanotube based multi-electrode arrays. *Nanotechnology*, **18**, 035201.

45 Ben-Jacob, E. and Hanein, Y. (2008) Carbon nanotube micro-electrodes for neuronal interfacing. *J. Mater. Chem.*, **18**, 5181–5186.

46 Shoval, A., Adams, C., David-Pur, M., Shein, M., Hanein, Y., and Sernagor, E. (2009) Carbon nanotube electrodes for

effective interfacing with retinal tissue. *Front. Neuroeng.*, **2**, 1–8.

47 Gabriel, G., Gómez, R., Bongard, M., Benito, N., Fernández, E., and Villa, R. (2009) Easily made single-walled carbon nanotube surface microelectrodes for neuronal applications. *Biosens. Bioelectron.*, **24**, 1942–1948.

48 Fan, S., Chapline, M.G., Franklin, N.R., Tombler, T.W., Cassell, A.M., and Dai, H. (1999) Self-oriented regular arrays of carbon nanotubes and their field emission properties. *Science*, **283**, 512–514.

49 Lin, C.M., Lee, Y.T., Yeh, S.-R., and Fang, W. (2009) Flexible carbon nanotubes electrode for neural recording. *Biosens. Bioelectron.*, **24**, 2791–2797.

50 Kim, D., Richardson-Burns, S.M., Hendricks, J.L., Sequera, C., and Martin, D.C. (2007) Effect of immobilized nerve growth factor on conductive polymers: electrical properties and cellular response. *Adv. Funct. Mater.*, **17**, 79–86.

51 Abidian, M.R. and Martin, D.C. (2009) Multifunctional nanobiomaterials for neural interfaces. *Adv. Funct. Mater.*, **19**, 573–585.

52 Abidian, M.R. and Martin, D.C. (2008) Experimental and theoretical characterization of implantable neural microelectrodes modified with conducting polymer nanotubes. *Biomaterials*, **29**, 1273–1283.

53 Qiao, Y., Chen, J., Guo, X.L., Cantrell, D., Ruoff, R., and Troy, J. (2005) Fabrication of nanoelectrodes for neurophysiology: cathodic electrophoretic paint insulation and focused ion beam milling. *Nanotechnology*, **16**, 1598–1602.

54 Kawano, T., Cho, C.Y., and Lin, L. (2007) Carbon nanotube-based nanoprobe electrode. Proceedings of the 2nd IEEE International Conference on Nano/Micro Engineered and Molecular Systems, pp. 895–898.

7
Hybrid Systems Analysis: Real-Time Systems for Design and Prototyping of Neural Interfaces and Prostheses

William Barnett and Gennady Cymbalyuk

7.1
Introduction

Neuroprosthetics is a frontier field at the intersection of neuroscience and biomedical engineering. It has proven itself efficient in restoring hearing, sight, and mobility to people challenged by disabilities [1–4]. Its success has fostered a strong demand for a variety of tools for design and prototyping so that neuroprostheses can be manufactured at the industrial scale necessary for practical medical applications. The modern technology of real-time systems is commonly used for development and prototyping in industries such as automotive, robotics, and aerospace. In neuroscience, real-time systems form the basis for such techniques as dynamic clamp and hybrid systems analysis see (see Chapter 5) [5–9]. Hybrid systems assembled from artificial and living neurons are instrumental in developing prototype neuroprostheses.

A neuroprosthesis implements and enables artificial neurons to interact with living neural circuits in order to provide functionality to network dynamics. In order for a neuroprosthesis to operate successfully, it must be biocompatible. Biocompatibility is often discussed in terms of the material composition of a neuronal interface, but here we emphasize biocompatibility in the sense that the dynamics of the artificial neuron do not compromise the functionality of network activity. Hybrid systems can be used to prototype for biocompatibility in neuroprostheses.

A hybrid system can be envisaged as a dynamical system. The activity of the system is determined by the biophysical characteristics, that is, parameters. We use bifurcation theory to find and manipulate control parameters in order to avoid undesirable dynamics brought on by such changes. Bifurcation theory allows us to identify and characterize equilibria and oscillatory regimes such as spiking and bursting activity in neuronal dynamics [10–12]. With it, we are also able to identify dynamical transitions or bifurcations. A bifurcation dictates quantitative laws describing the dynamics of the system. By applying these laws to a neuronal model or hybrid system, we can design artificial neurons with robust activity. For a range of parameter values in the artificial neuron, the activity of the hybrid system may appear functional even

Biohybrid Systems: Nerves, Interfaces, and Machines, First Edition. Edited by Ranu Jung.

though it is not robust. These regimes of activity are structurally not stable: small changes to biophysical parameters result in qualitatively different dynamics [13, 14]. Structural instability of artificial neurons can cause hazardous dynamics. Hazardous dynamics are regimes of activity that are susceptible to unwanted dynamical transitions. We consider two general cases of hazardous dynamics that are associated with bifurcation: catastrophes and multistability.

A catastrophe is an abrupt change in the activity of the system that occurs at the onset of some bifurcation. As a parameter changes in a way that approaches a bifurcation, the activity of the system changes smoothly. At a critical parameter value, the bifurcation occurs, and the activity of the system changes dramatically. A catastrophe is usually associated with multistability, which is the coexistence of attracting regimes. An external perturbation can then switch the system from one basin of attraction to another. Multistability has been observed in both the model and the living squid giant axon [15–17]. Environmental noise can cause the multistable neuron to switch back and forth between regimes irregularly [17].

In a neuron, the types of regimes that manifest as a result of catastrophe or multistability can be nonfunctional modes of activity. The functionality of an artificial neuron is critical for biocompatibility. We use bifurcation theory to identify hazardous regimes of activity that may compromise biocompatibility and to identify control strategies to maintain robust and functional dynamics. By investigating the dynamics of the model, we identify different regimes of activity spanning the high dimensional parameter space typical in a Hodgkin–Huxley style neuronal model. We can identify catastrophes and multistabilities that may dominate the activity of the system or are associated with hazardous dynamics. After any drawbacks and catastrophes of the hybrid have been studied and classified, a new generation of the prototype is to be designed and tested. This prototyping test phase is to be iterated until it achieves a hybrid system that is deemed biocompatible in this sense.

Hybrid systems combine the parameter controllability of mathematical modeling with the physical realism of electrophysiological experiments. From a technological point of view, the implementation of hybrid systems is facilitated by developments in dynamic clamp and real-time computing (Chapter 5). These hybrid systems require software operating in real time on a timescale that is fast compared to the fundamental time constants of any real or artificial ionic currents involved. Designated controllers are a proven means to fulfill these real-time requirements. A controller is a computer that performs computational tasks on a strict schedule. Using such a system to control a dynamic clamp enables real-time interaction between simulated and biological cells. Dynamic clamp has been used to implement artificial ionic currents in, and synapses between, living cells (Chapter 5) [5–9]. In hybrid systems, dynamic clamp provides the interface between the simulated and the living neurons [18]. These real-time solutions are often made from scratch. We use industrial off-the-shelf dSPACE real-time controllers in conjunction with the MATLAB and Simulink Real-Time Workshop to design and implement artificial neurons [19, 20].

In this chapter, we describe an easy-to-learn and -use technique with which to perform hybrid systems analysis. We describe a library of Simulink functions that are compatible with dSPACE controllers and sufficient to implement a typical dynamic clamp and to construct any Hodgkin–Huxley style model of a neuron (which is available at http://neuro.ni.gsu.edu/~gcymbalyuk/). Then, we provide an example of a real-time implementation of the Hodgkin–Huxley model of the squid giant axon and demonstrate hazardous dynamics. Our final example describes hybrid systems analysis of the leech heart interneuron and a half-center oscillator. We discuss the applicability of our tool for prototyping and interfacing biocompatible neuroprostheses.

7.2 Technology

The automotive, aerospace, and robotics industries rely on the extensive prototyping of control systems using real-time technology. The same technologies can be used for the development and testing of neuroprostheses by using hybrid systems analysis. A number of dynamic clamp solutions exist that satisfy the requirements for real-time computation (see Chapter 5). Real-Time Linux Dynamic Clamp and now Real-Time eXperiment Interface (RTXI; www.rtxi.org) are both open source platforms based on the real-time kernel extension for the Linux operating system [6]. National Instruments (Austin, TX) produces several lines of real-time PXI data acquisition boards in addition to the LabVIEW software suite that has been utilized for dynamic clamp [21]. Milescu *et al.* have created a dynamic clamp extension for the QuB program (www.qub.buffalo.edu) [22].

Hybrid systems need well-defined control systems and a robust platform in order to operate effectively in real time. Our models are built in Matlab and Simulink, which are a programming environment for technical computing (The MathWorks, Inc., Natick, MA). Simulink has a graphically represented language with extensive libraries of functions for common mathematical and analytical tasks. Some of these libraries, such as Real-Time Workshop, are dedicated to real-time and hardware-targeted computing. We use dSPACE real-time boards: the DS1104 R&D and the DS1103 PPC. dSPACE, Inc. (Paderborn, Germany) produces and distributes a series of control system-oriented real-time boards. A software package is bundled together with the dSPACE boards. It includes Real-Time Interface (RTI) and ControlDesk. RTI automatically converts block diagrams developed in Simulink into real-time code for dSPACE hardware. ControlDesk permits the building of a graphical user interface with elaborate instrument panels that can be connected with corresponding model variables for interactive control of real-time applications. dSPACE also provides a number of Simulink libraries for the design of control systems intended specifically for dSPACE hardware.

Our Simulink library is easy to use. Matlab and Simulink and the dSPACE hardware are over-the-counter and compatible with modern desktop computers. In addition, LabVIEW provides a number of tool-kits that allow Simulink models to be configured onto a number of National Instruments hardware targets. Our ready-made templates for Hodgkin–Huxley-type model components can be configured for a specific biological system and directly executed in real time.

7.2.1 dSPACE Boards

The dSPACE DS1104 R&D and DS1103 PPC controller boards are extensively used in the automotive, aeronautical, and robotics industries to prototype control systems. These controllers can be installed in most modern desktop computers. The DS1104 runs a Power PC 603e microprocessor at 250 MHz with 32 MB of SDRAM and 8 MB of flash memory. A 100-pin serial input/output ribbon connects the board to the CP1104 data acquisition board. The CP1104 has 16 BNC ports. The input voltage range is ± 10 V on eight analog-to-digital conversion BNC ports. The first four ports are multiplexed over one channel with 16 bits of resolution. Each of the four remaining ports has a dedicated channel with 12 bits of resolution. The eight output ports each have a range of ± 10 V with 16 bits of resolution.

The DS1103 is a full-sized ISA card. Alternatively, it can be housed in an external box with a PCI card adapter for the host PC. This larger board is equipped with a Power PC 750GX microprocessor running at 1 GHz with 32 MB local and 96 MB global SDRAM. Data acquisition is performed on the CP1103 board that links to the DS1103 via 3 parallel 100-pin serial ribbons. There are 32 BNC ports that handle input/output over the range of ± 10 V with 16 bits of resolution. Sixteen of the ports are analog-to-digital input ports multiplexed over four channels. Each of the eight digital-to-analog output ports has a dedicated channel.

While the dSPACE boards are primarily used to design and test airplane and car control systems, they are also efficient controllers for prototyping hybrid systems. Hardware specifications allow a board to simulate model neurons with the speed and fidelity necessary for electrophysiological protocols. When paired with standard laboratory equipment such as the AxoClamp-2B (Axon Instruments, CA), a Simulink model can efficiently run dynamic clamp in real-time on the DS1104 or DS1103.

7.2.2 Introduction to Programming in Simulink

Our dynamic clamp is implemented as a Simulink block diagram. We describe a library of functions that can be used to tailor it to specific needs. We have included a model of an oscillatory leech heart interneuron as well as basic functions for the construction of Hodgkin–Huxley-type neuronal models and functions to implement all standard dynamic clamp features.

Simulink is a graphical programming language for designing control systems in the MatLab programming environment. Operations and functions are represented by individual blocks or groups of blocks, and the flow of the program is made explicit by arrowed lines connecting each block. Blocks have input ports, where arguments and parameters are passed into the block, and output ports where the block returns the results of the operations or functions that it represents. Depending on the type of block, double clicking on a block allows the user to access its properties and options, embedded code, or the embedded subsystem of blocks. Simulink comes with 15 basic libraries of blocks in addition to several specialized libraries. The basic libraries contain blocks for math functions and signal processing, while the specialized libraries are targeted for more specific applications such as real-time control.

A small set of basic blocks is sufficient to create a neuronal model based on the Hodgkin–Huxley formalism (Figure 7.1a). These blocks work together when the output port of one block is connected to the input port of the next block. Each of these blocks has a number of ports, modes, and functions that can be changed by the developer. By configuring the internal settings of a block, the user can change the type of operation the block performs. For example, the Constant block supplies an unchanging signal of amplitude specified by the user. This signal may be a scalar, vector, or matrix depending on the configuration. The Gain block multiplies its input signal by a factor specified by the user. This can be used in place of the Multiplication/Division block. The Multiplication/Division block is versatile: the number and type of inputs can be tailored to the task at hand. It accepts scalar, vector, or matrix input. It is also possible to specify element wise or matrix operations. Setting this block with a single port that performs division is equivalent to obtaining the reciprocal for scalar inputs or performing matrix inversion on matrix inputs. Similarly, the Add/Subtract block has variable input ports and acts on scalar, vector, or matrix inputs. The Integration block performs numerical integration. There are both fixed step and variable step solvers available to the developer. The initial conditions for integration can be defined as a parameter internal to the block or they can be passed in as an argument. The Math Function block can be set to compute exponents, natural logarithms, powers of 10, logarithms of base 10, complex amplitude, squares, square roots, powers defined by input, complex conjugation, reciprocals, hypotenuses,

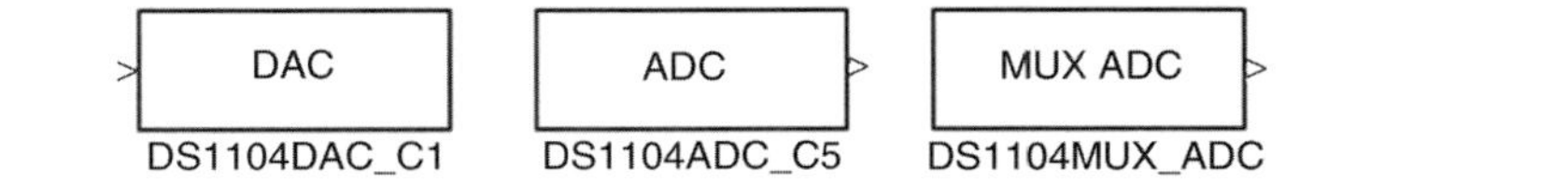

Figure 7.1 In Simulink, programs are represented as diagrams built out of elemental blocks. Each block represents an individual operation. (a) A small number of blocks representing basic mathematical functions are sufficient to create Hodgkin–Huxley-style neuronal models. (b) dSPACE blocks provide Simulink models with a means to interact with real-world systems. (b1,2) Channels with dedicated BNC ports can be accessed with individual blocks. The number at the end of the block name corresponds to the channel number of the port on the CP1104 board. (b3) The 16-bit analog-to-digital conversion channels are multiplexed. Internal block settings allow the developer to specify to which ports the block is connected.

remainders, moduli, transposes, and complex conjugate transposes. The Trigonometric Function block can perform functions including sine, cosine, tangent, arcsine, arccosine, arctangent, hyperbolic sine, hyperbolic cosine, hyperbolic tangent, and hyperbolic arctangent.

These Simulink blocks are sufficient to create sophisticated neuronal models, but do not provide us with a tool to interact with external systems. In order for a model to be useful for hybrid systems, a control signal from the system must be passed to an amplifier and into a cell as current, and the amplifier must feed the membrane potential of the cell back to the model. We move the simulation of the system off the desktop computer and onto a peripheral dSPACE board specialized for control systems. dSPACE has provided a set of Simulink blocks for writing data to and reading data from devices such as an amplifier. By passing the desired control signal from a model into a dSPACE block, the signal is routed out through the data acquisition board to the amplifier. Similarly, a dSPACE block acts as a source in a Simulink block diagram by routing the membrane potential of the cell back onto the control board. Blocks compatible with the DS1104 board are located in the DS1104 MASTER PPC library that is located in the dSPACE RTI1104 PPC directory in the Simulink library browser. dSPACE blocks allow the control system to interact with other devices via specific ports on the CP1104 (Figure 7.1b). The DS1104ADC_C5 block reads input from the board with 12-bit resolution. The C5 suffix indicates that the block corresponds to port 5, but the block is applicable for ADC ports 5 through 8. The DS1104DAC_C1 block writes to a digital-to-analog port with 16-bit resolution and is applicable for all eight digital-to-analog ports. For the DS1104DAC and DS1104ADC blocks, the user can specify the channel with which it is associated by changing the number in the suffix to the desired port number, provided that it is within the applicable range. Finally, the DS1104MUX_ADC block reads from analog-to-digital ports 1 through 4 at 16 bits. In order to specify which port or ports, it is necessary to change the settings of the block.

7.2.3 Library for Dynamic Clamp

The dynamic clamp is a tool used to perform electrophysiological experiments by injecting current with particular dynamical characteristics into a living cell. In this manner, an artificial current with characteristics defined by the researcher can be introduced in addition to the existing set of ionic currents endogenous to the membrane. Alternatively, the dynamic clamp can be used to substitute an artificial current with desirable characteristics for an endogenous ionic current. A hybrid system uses the dynamic clamp as a link between the computational and the living components of the system in order to realize real-time control of the temporal characteristics of neuronal dynamics. The dynamic clamp is described in detail in Chapter 5.

Let us consider the hyperpolarization-activated current (I_h) identified and measured in the heart interneuron (HN) of the medicinal leech and then implemented in the canonical model of HN [23, 24]. The membrane potential and time in this model of the HN are in units of volts and seconds. The equations for the current and its activation are

$$m_{h\infty}(V) = \frac{1}{1 + 2\,\exp(180[V + 0.047]) + \exp(500[V + 0.047])},$$

$$\tau(V) = 0.7 + \frac{1.7}{1 + \exp(-100[V + 0.073])},$$

$$\frac{dm_h}{dt} = \frac{m_{h\infty}(V) - m_h}{\tau(V)},$$

$$I_h = \bar{g}_h m_h^2 [V - E_h]$$

such that $\bar{g}_h$ is the maximal conductance, E_h is the reversal potential, and m_h is the activation variable. The functions $m_{h\infty}(V)$ and $\tau(V)$ are the steady state and time constant of activation. These types of models lend themselves to hierarchical organization. At the highest level, the model, environmental parameters, and dSPACE links to the data acquisition board are assigned. Descending into the model reveals a menagerie of blocks and subsystems.

A block diagram built to perform a specific task can be condensed into a single block so that complicated systems can be organized into a set of subsystems. A subsystem is a block diagram that is represented by a single block. Subsystems make it easy to conceptualize and organize the flow of operations in a block diagram by establishing a tree-like hierarchy for functional model components. Mundane details of computation can be sequestered to lower hierarchical levels while control parameters can be readily available for manipulation closer to the root of the model.

At the lowest organizational level, system functions are evaluated. The operations performed at this level are primarily the calculations of steady-state membrane potentials and gating variable time constants. The steady-state membrane potential of I_h is a sigmoidal function (represented by $f(b, V)$ in Figure 7.2a). The corresponding block diagram is inserted into the subsystem block hnInfh. The parameter b is an argument of the function and an input port of the block. It will be assigned by a Constant block at a higher organizational level. Other parameters of the function, such as the Gain blocks set at 180 and 500, can still be accessed from ControlDesk in real time, but are left at this lower organizational level because there is no anticipation of accessing them. The steady-state membrane potential for the activation of I_h is a unique function. The function for the voltage-dependent time constant of activation, however, is frequently used in our model. From current to current, these functions are represented by similar equations differing only in parameter values. Rather than build a dozen unique diagrams for time constant functions, a small set of equations can be reused to define many gating variables. A common equation for a voltage-dependent time constant is also a sigmoidal function and contained in the subsystem block hnTau (represented by $f(a, b, c, d, V)$ in Figure 7.2b). In this case, each of the parameters is assigned at a higher organizational level so that the same block can be used in more than one gating variable.

The equation for the derivative of the activation of I_h, dm_h/dt is presented as a block diagram in Figure 7.3a. The Simulink subsystem block for this equation is mh with

$$f(b, V) = \frac{1}{1 + 2 \cdot \exp(180) \cdot (b + V)) + \exp(500 \cdot (b + V))}$$

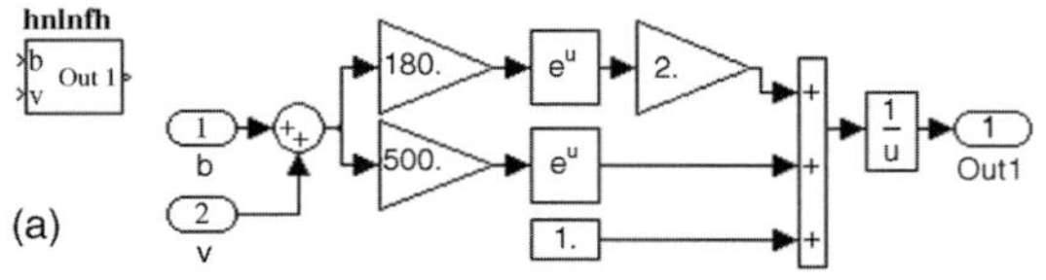

$$f(a, b, c, d, V) = c + \frac{d}{1 + \exp(a \cdot (b + V))}$$

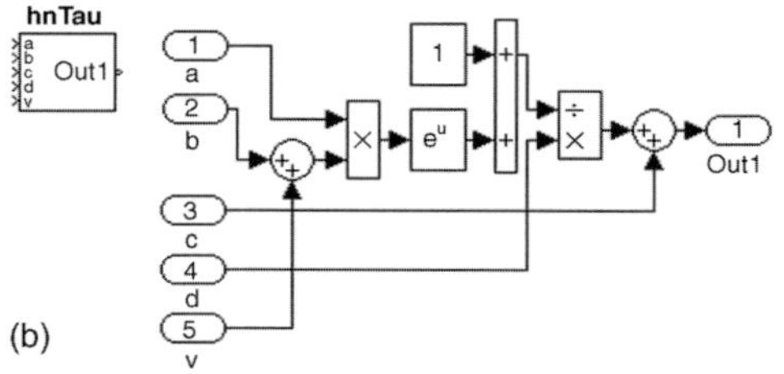

Figure 7.2 (a) The function $f(b, V)$ denotes the steady-state membrane potential of I_h. The block diagram equivalent to $f(b, V)$ is embedded in the subsystem block hnInfh. In this case, a single parameter is defined at a higher organizational level. (b) The voltage-dependent time constant of activation of I_h can be written as $f(a, b, c, d, V)$, and its block diagram is embedded in the subsystem block hnTau. This function is not unique to I_h. By defining many parameters at a higher organizational level, this block can be used for different time constants in other subsystems of the same model.

one input port, V, for the membrane potential and one output port, mh, for the value of the gating variable. Inside the subsystem, the dynamics of the gating variable are defined by the steady-state activation block hnInfh and the time constant block hnTau. At this hierarchical level, the parameters left undefined inside these blocks are passed in as arguments along with the membrane potential. It becomes important, when working in ControlDesk, to give these Constant blocks names that explicitly identify not only the parameter that each represents but also the specific current, variable, and function with which each is associated so that they may be differentiated from similar blocks in other subsystems.

Each ionic current subsystem has input ports for the membrane potential, the maximal conductance, and the reversal potential and output ports for the current and each of the intrinsic gating variables (Figure 7.3b). The eight subsystem blocks representing the voltage-gated currents are assembled to create a complete model of the HN neuron. This model is implemented as the “HN Neuron” subsystem (Figure 7.4).

The mutually inhibitory HN cells form a robust half-center oscillator. Our model utilizes a fast spike-mediated synapse as the agent of this inhibition [24, 25]. The activation and modulation of I_{Syn} are dynamical variables that are controlled by the presynaptic membrane potential. Spikes of the presynaptic cell lead to an increase in

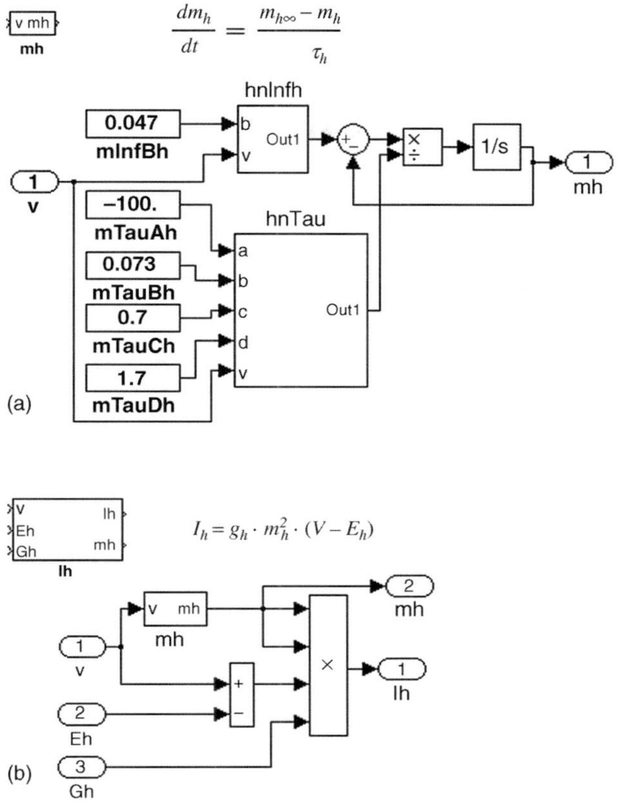

Figure 7.3 (a) The activation variable for I_h is determined by the equation for dm_h/dt that can be put together using preexisting functions for steady state of activation and time constant of activation (see Figure 7.5). At this hierarchical level, constants left undeclared in constituent functions are assigned and passed into the appropriate input ports. (b) The current subsystem, Ih, has two outputs: values for m_h and I_h. Functionally, the block is meant to evaluate the current, but it is convenient to pass out the activation variable as well, so that its dynamics can be recorded for further analysis.

the conductance of the current, and the membrane potential of the postsynaptic cell drives the synaptic current. The synaptic current on the postsynaptic cell is summed and integrated along with the intrinsic ionic currents of the membrane. The updated membrane potential is fed back into the intrinsic ionic current subsystems and is also passed out of the cell to determine the activation of the synapse onto the other cell. The neuronal subsystem HNwFastSyn implements the synaptic current by including the ISynS ionic current block and an additional input port for the presynaptic membrane potential.

A hybrid system has both a living component and a computational component. An electrode placed inside of a living neuron reads the membrane potential and injects

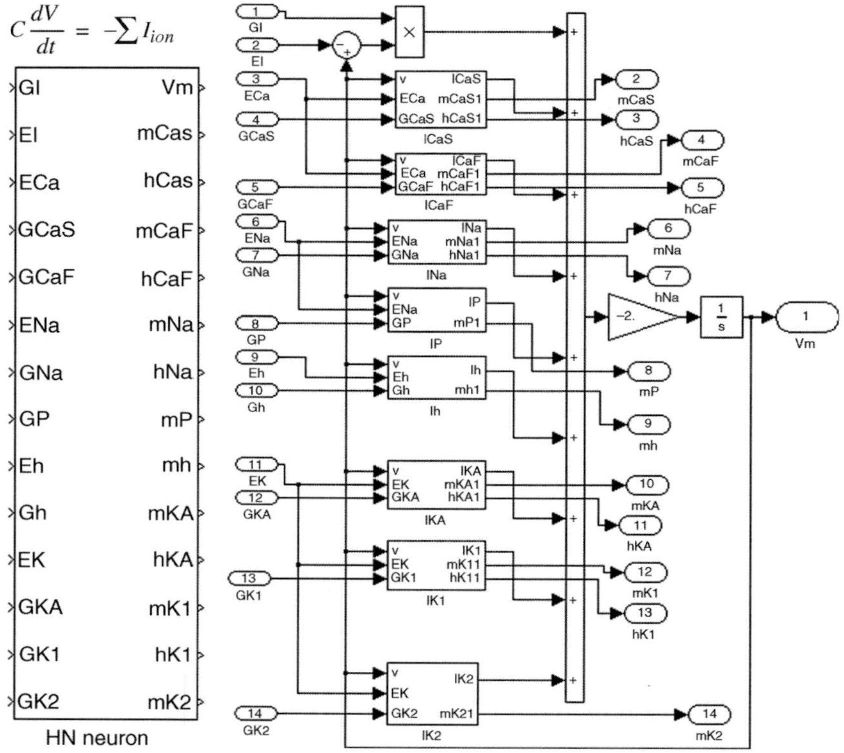

Figure 7.4 This is the model hierarchical level. Each of the eight voltage-gated currents is implemented as a subsystem block to keep the flow of the program uncluttered and allow maximum utilization of generic library functions. Conductance and reversal potential parameters are assigned at and each of the state variables is passed out to a higher organizational level for ease of access in ControlDesk.

current. Each of these processes is mediated by an amplifier and a data acquisition board. A procedure running in real-time allows a researcher to analyze captured data and directly control the dynamics of an injected current on the fly. By implementing our neuronal model in such a procedure, the model may interact with a living cell. The membrane potential of the living cell is read off the electrode and passed to the model running in real time. This data feed is utilized to produce a synaptic response on the model neuron, and the response of the model cell is used to calculate a synaptic current that is then injected back into the living cell. The subsystem block MathAndLive demonstrates an implementation of just such a procedure on the DS1104 controller.

By connecting the voltage out and injected current command ports on an amplifier to the ADC and DAC ports on the CP1104 data acquisition board, the dSPACE ADC and DAC Simulink blocks provide a neuronal model direct access to its living counterpart. The membrane potential of the living cell is passed from the ADC block to the model cell where the live-to-math synapse is calculated. This current is summed with the intrinsic currents of the model neuron and integrated. Meanwhile, the membrane potential of the model cell is used to calculate the activation of the math-to-live synapse, and the membrane potential of the living cell is used to calculate the synaptic current. The value of the current is passed into the DAC block, and the signal is fed from the CP1104 board into the injected current command port on the amplifier.

7.3 Applications

7.3.1 Building Neuronal Models with Simulink for Real-Time Analysis

Consider the model of the squid giant axon [26]. It includes a fast sodium current (I_{Na}), a delayed rectifier potassium current (I_{K}), a leak current (I_{L}), and an injected current (I_{inj}). The differential equation for membrane potential takes the form

$$\frac{dV}{dt} = \frac{1}{C}\left[I_{\text{inj}} - \bar{g}_{\text{K}} n^4 [V - E_{\text{K}}] - \bar{g}_{\text{Na}} m^3 h [V - E_{\text{Na}}] - \bar{g}_{\text{L}} [V - E_{\text{L}}]\right],$$

where the gating variables are defined as the differential equations

$$\frac{dn}{dt} = \alpha_n [1-n] - \beta_n n,$$

$$\frac{dm}{dt} = \alpha_m [1-m] - \beta_m m,$$

$$\frac{dh}{dt} = \alpha_h [1-h] - \beta_h h$$

and the rate functions are

$$\alpha_n = \frac{10 \cdot [10 - V]}{\exp([10 - V]/10) - 1},$$

$$\beta_n = 125 \cdot \exp(-V/80),$$

$$\alpha_m = \frac{100 \cdot [25 - V]}{\exp([25 - V]/10) - 1},$$

$$\beta_m = 4000 \cdot \exp(-V/18),$$

$$\alpha_h = 70 \cdot \exp(-V/20),$$

$$\beta_h = \frac{1000}{\exp([30 - V]/10) + 1}$$

such that V is the membrane potential in millivolts, n is the activation of the potassium current, m is the activation of the sodium current, and h is the inactivation of the sodium current. In this model, time is in seconds. The maximal conductance of an ionic current and its reversal potential are $\bar{g}_{\text{ion}}$ and E_{ion}. The parameter values used for maximal conductances are $\bar{g}_{\text{K}} = 36\,\text{mS/cm}^2$, $\bar{g}_{\text{Na}} = 120\,\text{mS/cm}^2$, $\bar{g}_{\text{L}} = 0.3\,\text{mS/cm}^2$, and the values used for reversal potentials are $E_{\text{K}} = -12\,\text{mV}$, $E_{\text{Na}} = 115\,\text{mV}$, and $E_{\text{L}} = 10.13\,\text{mV}$.

Hodgkin–Huxley style neuronal models typically share a common structure. Gating variables are defined by voltage-dependent rate functions, currents are defined by gating variables, and the membrane potential is computed by integrating the sum of the ionic currents. The equations at each level of a model are often identical apart from parameter values. We make Simulink block diagrams as generic as possible, so that the same block can be used to build more than one gating variable or ionic current. With this copy-and-paste style of programming, it is easy to compile libraries of multipurpose functions.

The dynamics of the squid giant axon model are determined by three channel opening rates and three channel closing rates. Each of these six rate functions can be described by one of the following three equations:

$$k_1(c, d, V) = \frac{c[d-V]}{\exp([d-V]/10)-1},$$

$$k_2(c, d, V) = c \cdot \exp(-V/d)x,$$

$$k_3(c, d, V) = \frac{c}{\exp([d-V]/10)+1}.$$

Functions $k_1(c, d, V)$, $k_2(c, d, V)$, and $k_3(c, d, V)$ are embedded in the subsystem blocks k_1, k_2, and k_3 with input ports for c, d, and the membrane potential (Figure 7.5). Block k_1 corresponds to rates α_n and α_m; block k_2 corresponds to rates β_n, β_m, and α_h, and block k_3 corresponds to the rate β_h.

These rate functions and their associated parameters identify the otherwise generic gating variable derivative blocks. The gating variable blocks are used in much the same way; once the correct rate function and parameters are assigned, the derivative block can be dropped directly into the model. For example, the block dn%dt computes $\mathrm{d}n/\mathrm{d}t = \alpha_n[1-n]-\beta_n n$. At this level of organization, the derivative block calls the rate function blocks k_1 and k_2 and assigns the rate parameters for α_n and β_n (Figure 7.6). It uses the current gating variable and membrane potential value to evaluate these functions and outputs the derivative of the gating variable with respect to time. The equations governing the conductances of the ionic currents are largely unique, so there is a dedicated block for each current.

In this model, integration is performed as a vector operation. The derivatives of each of the variables of state are multiplexed into a single signal and passed into the integrator. In the block diagram, the multiplexed signal is a thicker connecting arrow (Figure 7.7). The actual operation of integration is performed on each derivative independently. There are many methods for numerical integration. The simplest method is Euler's algorithm where, given some initial value of a function, F, the

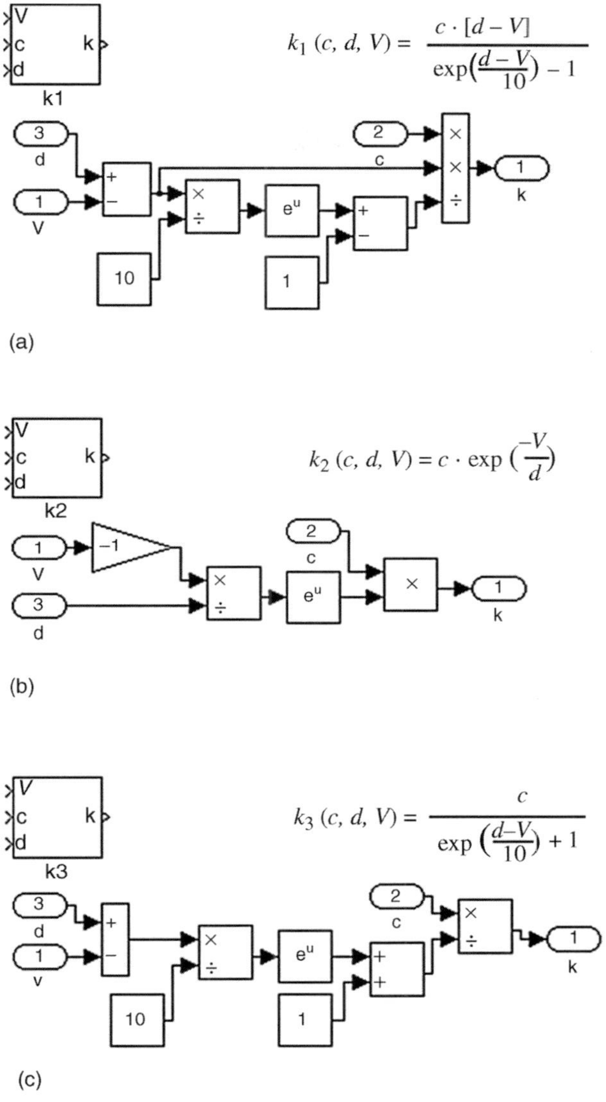

Figure 7.5 With these three unique subsystem blocks, one can implement in Simulink any of the six rate equations of the Hodgkin–Huxley model. (a) The block k_1 represents function $k_1(c, d, V)$. This block is used for the rates α_n and α_m. (b) Block k_2 represents function $k_2(c, d, V)$ that corresponds to rates β_n, β_m, and α_h. (c) Subsystem block k_3 represents the function $k_3(c, d, V)$ that corresponds to β_h.

derivative of the function is evaluated at discrete intervals of time in order to produce a numerical approximation of F as it evolves over time. Other methods may produce a more accurate approximation but may not meet the time constraints imposed by the application.

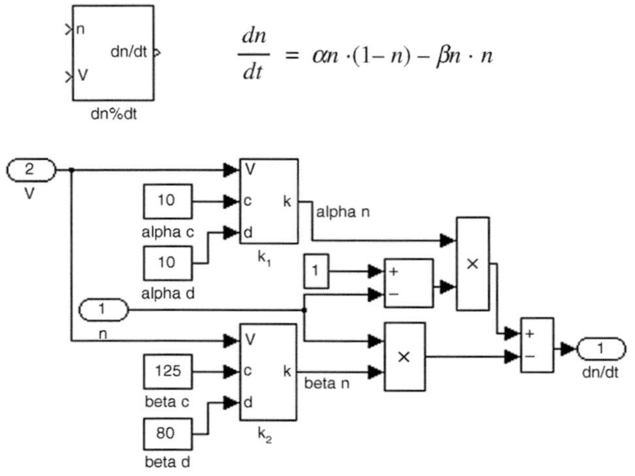

Figure 7.6 The maximal conductance of a current is determined by the component gating variables. The temporal characteristics of each of these gating variables are determined by a pair of rate functions. The rate function blocks are inserted into a generic derivative block, and the parameter inputs are specified. Take, for example, the $dn\%dt$ block. Blocks k_1 and k_2 are the appropriate rate equations for α_n and β_n.

7.3.2
Propensity to Hazardous Dynamics of the Squid Giant Axon

For the parameter values provided by Hodgkin and Huxley, this model does not exhibit periodic tonic spiking. A brief depolarizing injected current pulse may elicit an individual spike of the membrane potential, and a constant depolarizing injected current may bring the cell into a regime of tonic spiking. For a particular range of values for I_{inj}, there is coexistence of a stable stationary state and a stable tonic spiking regime [15, 16]. These two regimes are separated by the stable manifold of a saddle orbit. The coexistence means that the neuron could be found in either the silent or spiking regime. Also, it has been shown that a pulse of current can switch the activity of the neuron between these two regimes [15, 16].

This scenario can be tested by applying episodes of noise to the artificial cell (Figure 7.8) [17]. We introduce a mechanism through which a model can interact with external processes. For this example, the dSPACE DS1104 hosts the model of the squid giant axon. The basic Simulink and dSPACE data acquisition blocks may all be compiled to run in real-time. The DS1104MUXADC block reads input from the first four analog-to-digital channels on the CP1104. This block will read the external stimulus used to switch the model from spiking to silence. The ADC block feeds directly into a gain block, so that the stimulus signal can be amplified to sufficient magnitude and is also modified by an offset current to keep the ambient stimulus signal at zero.

The model is compiled with Real-Time Workshop and is loaded onto the dSPACE board. We have designed an example of a graphical user interface implemented in

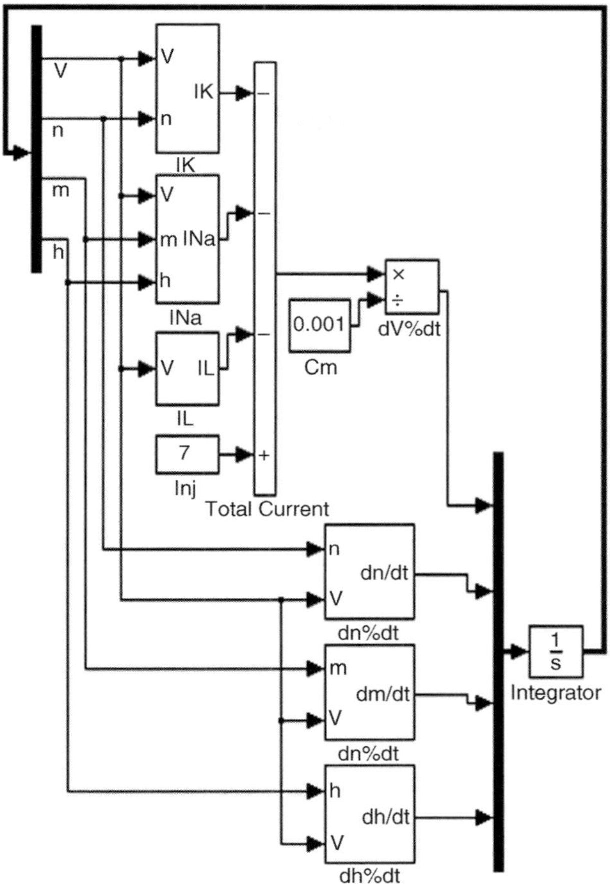

Figure 7.7 The Hodgkin–Huxley model of the squid giant axon implemented as a block diagram. The variable derivatives are multiplexed onto one signal, and the system is integrated as a vector. The resultant state variables are demultiplexed and passed into the system. Rather than multiple Integrator blocks operating at different hierarchical levels, integration is carried out at the model level of organization. Dark upright blocks are multiplexers and demultiplexers. Thick connecting arrows indicate multidimensional signals.

ControlDesk, which contains tools to interact with and control the model in real time. It includes a plot and control panel to monitor and capture the membrane potential and stimulus current. The layout also includes two numerical input fields to control the amplitude of the stimulus current and the offset current. It may be necessary to zero the ambient stimulus current by adjusting the offset block so that a baseline stimulus current does not modify the dynamics of the cell.

To provide an external stimulus, an operator touched the ADC1 port on the CP1104 with a conducting pin. This procedure provided a 60 Hz noisy stimulus. The voltage

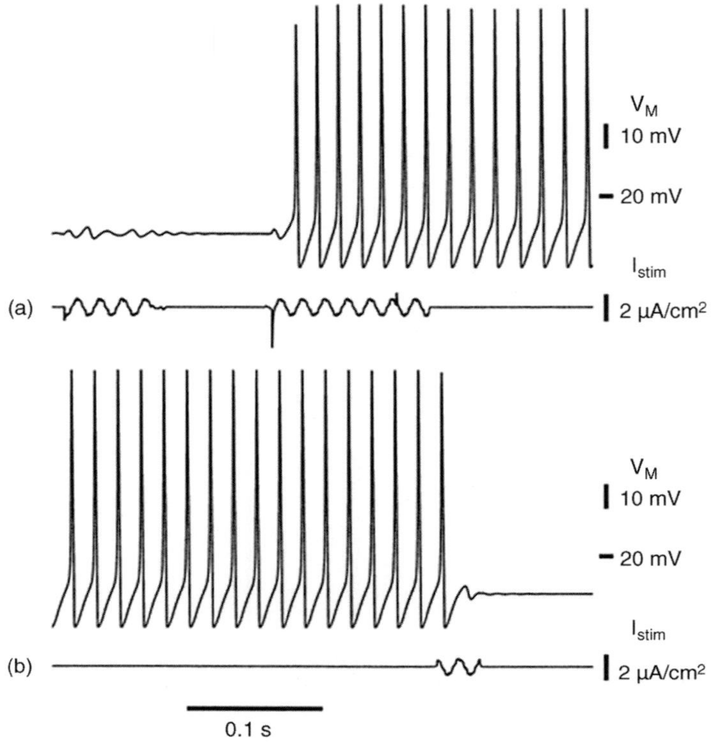

Figure 7.8 A noisy 60 Hz stimulus current may cause a switch from silence to tonic spiking or cause the annihilation of tonic spiking. This current was generated when the operator touched the input port of the data acquisition board. (a) A stimulus is insufficient to cause a switch from silence to spiking. The second stimulus does switch the model cell from a silent to a spiking mode. (b) A brief stimulus terminates the spiking activity.

and current traces were monitored and recorded using ControlDesk (Figure 7.8). As long as the contact with the port is maintained, noisy excursions are observed in the stimulus current.

7.4 Hybrid Systems Analysis in the Leech Heart Interneuron

Hybrid systems analysis is a powerful technique that has proven itself in neuroscience. We show its usefulness as a prototyping tool for biomedical engineers. We consider the hybrid system to be a neural interface implemented with dynamic clamp. Previously, the technology has been used to identify the role of individual currents in specific cellular mechanisms. The intent of prototyping may be to establish a certain stable regime of circuit activity, but it is hybrid systems analysis that we use to reveal the effect of explicitly defined ionic currents of the artificial cell on the activity of a neural circuit. By replacing a component of a living circuit with a computational cell, we gain knowledge of its role in the circuit by manipulating the parameters of artificial ionic currents. With prototyping, we seek to take the knowledge of the role of individual ionic currents and move an unhealthy circuit from a pathological regime to a healthy regime. We present a set of experiments performed on the leech heart interneuron as an example of prototyping and hybrid systems analysis. The first study utilizes a VLSI silicon artificial neuron in which I_h is tuned to examine its interaction with the synaptic current [19]. The second study uses computational neurons built with the library we present in Section 7.2.3 to tune the slow Ca^{2+} current to show its role in supporting bursting activity [20].

7.4.1 Model Heart Interneuron

The artificial cell is composed of eight voltage gated currents and a leak current [24]. The eight voltage gated currents are composed of thirteen gating variables. The mathematical model is composed of 14 equations: one for the membrane potential and one each for the gating variables. The hybrid system contains a synapse from the artificial cell onto the living cell and a synapse from the living cell onto the artificial cell. The synaptic current has a spike-mediated component and a slower modulating component, so the system requires an additional six differential equations:

$$I_{\text{Syn}} = \bar{g}_{\text{Syn}} Y_{\text{post}} M_{\text{post}} [V_{\text{post}} - E_{\text{post}}],$$

$$\frac{dX_{\text{post}}}{dt} = \frac{[X_{\infty}(V_{\text{pre}}) - X_{\text{post}}]}{0.002},$$

$$\frac{dY_{\text{post}}}{dt} = \frac{[X_{\text{post}} - Y_{\text{post}}]}{0.011},$$

$$\frac{dM_{\text{post}}}{dt} = \frac{[M_{\infty}(V_{\text{pre}}) - M_{\text{post}}]}{0.2},$$

$$X_{\infty}(V_{\text{pre}}) = \frac{1}{1 + \exp(-1000[V_{\text{pre}} + 0.01])},$$

$$M_{\infty}(V_{\text{pre}}) = 0.1 + \frac{0.9}{1 + \exp(-1000[V_{\text{pre}} + 0.04])}.$$

The membrane potential of the living cell is recorded and passed into the model system where it is used to calculate the synaptic current received by the artificial cell. The membrane potential of the artificial cell is used to evaluate the synaptic current onto the living cell, which is passed out of the system to be injected by an amplifier into the live cell. We are able to integrate each of these operations with Euler's method for numerical integration at 10 to 20 kHz on the DS1103 PPC board. At these frequencies, the artificial system performs sufficiently fast to produce biophysically plausible dynamics.

7.4.2
Hybrid Systems Analysis

The heartbeat of the leech is controlled by a central pattern generator (CPG). The kernel of this CPG consists of pairs of bursting HN cells, which form half-center oscillators. In a half-center oscillator, HN cells inhibit each other with inhibitory cholinergic synaptic currents and burst in alternation [27]. On the inhibited neuron, the prolonged hyperpolarization during the interburst interval activates I_{h}, and the balance between I_{Syn} and I_{h} determines the duration of its interburst interval [19, 24]. Upon depolarization, the conductance of I_{CaS}, the slow low-threshold Ca^{2+} current, becomes large and then slowly dwindles as the current is inactivated [20]. This long-lasting inward current supports the plateau depolarization of the burst and regulates the spike frequency. The time course of inactivation of I_{CaS} determines the cell's burst duration [20, 24]. These three currents play key roles in determining the temporal characteristics of the leech heart half-center oscillator [19, 20, 24, 25].

Studies of the leech heart half-center oscillator pose a technical difficulty. Observations with sharp intracellular electrodes have shown that the HN cells spike tonically when pharmacologically isolated from the cholinergic synapse with bicuculline methiodide [27]. However, recordings with extracellular electrodes under the same pharmacological conditions show that HNs burst endogenously [25]. The model shows evidence that a nonspecific leak current, introduced by piercing the cell with a sharp electrode, is responsible for the onset of tonic spiking. In order to implement a hybrid system, a sharp electrode is used to inject an artificial current into the living cell. Even though the HN is an endogenously bursting cell, dynamic clamp experiments are performed while the cell is in a tonic spiking mode. Nevertheless, the inhibitory synaptic coupling between artificial and living HN cells is sufficient to establish antiphasic bursting activity [19, 20]. We tune the hybrid system by manipulating the parameters of the artificial cell to identify the mechanisms supporting the half-center oscillator.

In order to establish a biophysically plausible hybrid half-center oscillator, a certain amount of tuning is necessary. Key parameters for temporal regulation can be identified in an analysis of the parameter space ahead of time, such that the artificial cell can be adapted to its intended synaptic target quickly at run time [24, 25]. The uncoupled artificial cell must exhibit spiking activity of roughly the same frequency as its living counterpart that was achieved by tuning the leak current [20].

The experiments that ascertain the role of I_{h} and I_{CaS} encapsulate the essence of prototyping. Sorensen *et al.* show that by decreasing g_{h} in the artificial cell, the interburst interval of the artificial cell increases [19]. When g_{h} is increased in the artificial cell, the counteraction to inhibition increases and the interburst interval of the artificial cell decreases in duration. Olypher *et al.* describe the role of I_{CaS} with a similar set of experiments (Figure 7.9) [20]. By decreasing the time constant of inactivation of I_{CaS}, $\tau_{\mathrm{h,CaS}}$, the conductance of I_{CaS} is inactivated at a greater rate in

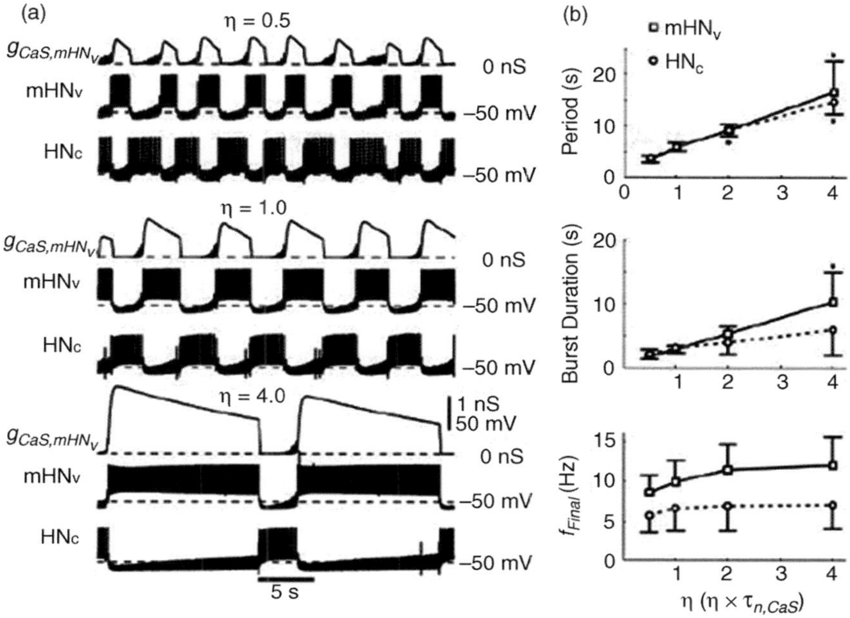

Figure 7.9 A hybrid system consists of a living cell coupled to a model cell via artificial ionic currents. g_{CaS} is the conductance of I_{CaS}. mHN_v is the model HN. HN_c is recorded from a living HN via sharp electrode. Removal of inactivation during the hyperpolarized quiescent phase of the burst supports the plateau depolarization during the active phase of the burst. The membrane potential of the model cell and both synaptic currents are calculated in real time. The membrane potential of the living cell is monitored via intracellular electrode. (a) Example trajectories showing activity for three values of η. (b) Period, burst duration, and final spike frequency for $\eta = 0.5$, $\eta = 1$, $\eta = 2$, and $\eta = 4$. Reproduced from Olypher, Cymbalyuk, Calabrese, J. Neurophysiology, 2006 [20] with permission from Am. Physiol. Soc.

response to the depolarization during the burst, so the bursts are shorter, and the interburst interval of the presynaptic HN is shortened. An increase in $\tau_{h,CaS}$ in the artificial cell allows I_{CaS} to stay activated for a longer period. The prolonged burst of the artificial cells leads to a prolonged synaptic current on the living HN cell.

7.5 Discussion

Neuroprosthetic devices will one day routinely treat neurological disorders by monitoring and regulating the electrical activity in the central nervous system. Such brain–machine interfaces will need to be small, fast, and power efficient. The introduction of the analog very large-scale integration (aVLSI) neuron provides a framework for artificial cells and neural networks that satisfies these criteria [28]. Silicon neurons working in tandem with living networks have produced biologically valid patterns [19, 29]. The optimization of circuit design promises efficient and biocompatible VLSI chips [30]. The clinical distribution of these neuroprostheses will require industrial manufacturing of implants. Like in any other industry, the neuronal models upon which these devices will be based will need to be tested before the production process begins. Testing could be done with a hybrid system: artificial neurons can be interfaced with biological neurons or networks of neurons with a dynamic clamp. By tuning characteristics of the artificial cell, an analysis of the hybrid system can reveal pathological or otherwise undesirable neuronal or network states. By prototyping neuroprostheses in this manner, a researcher can design an artificial cell that meets implant design parameters without compromising the dynamics of the biological component of the hybrid system.

In order for a hybrid system to function properly, the artificial system must be simulated on a computer such that the computational cell is able to respond to the living cell without interruption. These computational necessities are often accomplished with real-time computing in which operations are performed on a strict schedule. Developers must rely on specialized products to implement real-time applications: a real-time operating system runs a program directly on the desktop computer or a program is run on a real-time peripheral device (a control board) that is controlled from the desktop computer. A prominent supplier of real-time control boards and support software is dSPACE, Inc. dSPACE products are commonly used for prototyping control systems. dSPACE software automatically utilizes a system designed in Simulink and executes it on a real-time control board. We realize hybrid systems by designing neuronal models in Simulink and using dSPACE to implement these models in real time.

Matlab and Simulink provide a high-level programming language designed for technical and high-performance computing. Simulink is a graphical programming environment. Even an inexperienced user can go from a set of equations or a conceptual flow chart to a sophisticated model quickly with Simulink's drag-and-drop programming style. As in any high-level programming language, program components can be reused to reduce development time. We describe a library of functions in Simulink sufficient to create a neuronal model. This library includes individually implemented artificial currents, voltage-dependent steady-state curves, and voltage-dependent time constants. Moreover, these functions provide a template with which to construct novel neuronal models. Many of our functions are standardized so that each numerical parameter is specified by the user. By providing custom parameters based on experimentally measured ionic currents, a researcher could implement any number of currents not found in our library. These models are designed to be used as real-time control systems for a dynamic clamp.

In order for a neuroprosthesis to assume the role of some functional neuronal circuit, it must behave as though it is a functional neuronal circuit. That means that it must continue to have a functional role in its host as the local and global neuronal environments change over time. Factors such as fatigue, age, injury, and drug use can all change the way neuronal circuits and their synaptic targets function. Prosthetic activity must be robust so that the larger neuronal circuit remains resistant to degeneration. We use the numerical and analytical tools of dynamical systems theory to study the mechanisms through which these environmental factors enact neuromodulation in order to identify manifestations of pathology in model neurons.

Dynamical systems theory provides a theoretical foundation for sensitivity analysis in mathematical models. Qualitative aspects of a dynamical system can be revealed through mathematical analysis of critical parameter values and initial conditions in the context of perturbations and environmental noise. For example, there may exist multiple stable regimes nearby each other in state space (multistability), or the activity of a system may change dramatically with small changes to sensitive parameters (catastrophe). Regimes of activity that exist in a multistable or potentially catastrophic system could be hazardous if small perturbations to the system cause it to switch to a different mode of activity. Multistability and catastrophes are not mutually exclusive. A catastrophe may manifest itself as one of the many stable modes becomes unstable or disappears. Forearmed with the knowledge of the type of activity that can be expected across a range of parameters, a researcher may design a model neuron in order to exclude or take advantage of certain regimes.

Multistability and catastrophes can be used to describe several facets of intrinsic neuronal dynamics and modulation. A neuron may exhibit coexistence of stable regimes [16, 17, 31]. Such a cell could toggle between these regimes upon stimulation. Using dynamical systems theory, a researcher may predict such multistability in a neuronal model. If the model were intended to fire indefinitely as a pacemaking oscillator, then bistability would be a pathological trait. On the other hand, bistability could be a functional trait if the model is intended to be an on–off switch that modulates some macroscopic behavior. A model neuron may switch from endogenous bursting to tonic spiking at a particular parameter value [25, 31]. A researcher may implement this parameter as a slow variable dependent on the model's synaptic input or state of activity in order to effect the modulation of neuronal dynamics.

In this chapter, we described a technique for the implementation of hybrid systems based on the technology provided by dSPACE Inc. This technique has proven fruitful for analysis of dynamics of small neuronal networks [19, 20] as was described in more detail in Chapter 5. Concerning prototyping of prosthetic devices, we emphasized development of individual artificial neurons with biophysically accurate dynamics and focused this chapter on systems composed of individual artificial and living neurons. This approach, requiring successive iterations of development of artificial neurons and analysis of hybrid systems, uses intracellular electrodes and is not directly applicable to human subjects. We suggest that preliminary prototyping could be done on invertebrate neural circuits that are dynamically similar to target circuits in the human nervous system. For example, the stomatogastric CPG has been proposed as a conceptual model for cortical circuits (Chapter 5) [32].

Contemporary neuroprostheses deal with large numbers of neurons. These applications are based on extracellular electrodes, and dSPACE technology is readily compatible with the spike-driven signals that these prostheses record as input from the host. On the other hand, the technology providing feedback from the prostheses to living tissue is an open problem for bioengineering, although a number of promising technologies are under development. For example, microelectromechanical and microfluidic-based technologies have opened an interesting avenue for prosthetically controlled chemical input to neuronal or otherwise excitable tissue [33–36]. We have not described these technologies in detail because their implementation lies outside the scope of our tutorial on developing hybrid systems utilizing real-time controllers. However, future prostheses will combine extracellular recordings with chemical input to implement local and artificial synapses. We assert that these technologies will be driven by artificial neurons or networks of artificial neurons, and whether these computational models are biophysically accurate or phenomenological, such a hybrid system must be prototyped to ensure dynamical biocompatibility.

Acknowledgments

The development of the software presented here was supported by NIH NINDS grant NS43098 to Ronald Calabrese, Gennady Cymbalyuk, and Stephen DeWeerth. Most of these codes were developed in Ronald Calabrese's laboratory in Emory University. We also acknowledge support by NSF grant PHY-0750456 to Gennady Cymbalyuk. We would like to thank Ariel Hart, Ryan Hooper, Alison McCullough, Gabrielle O'Brien, and Astrid Prinz for carefully reading this chapter.

References

1 Moro, E. and Lang, A.E. (2006) Criteria for deep-brain stimulation in Parkinson's disease: review and analysis. *Expert Rev. Neurother.*, **6** (11), 1695–1705.

2 Ostrem, J.L. and Starr, P.A. (2008) Treatment of dystonia with deep brain stimulation. *Neurotherapeutics*, **5** (2), 320–330.

3 Ackermans, L., Temel, Y., and Visser-Vandewalle, V. (2008) Deep brain stimulation in Tourette's Syndrome. *Neurothereapeutics*, **5** (2), 339–344.

4 Fayad, J.N., Otto, S.R., Shannon, R.V., and Brackmann, D.E. (2008) Auditory prostheses "neural interface for hearing restoration: cochlear and brain stem implants." *Proc. IEEE*, **96** (7), 1085–1094.

5 Sharp, A.A., O'Neil, M.B., Abbott, L.F., and Marder, E. (1993) The dynamic clamp: artificial conductances in biological neurons. *Trends Neurosci.*, **16**, 389–394.

6 Dorval, A.D., Christini, D.J., and White, J.A. (2001) Real-time linux dynamic clamp: a fast and flexible way to construct virtual ion channels in living cells. *Ann. Biomed. Eng.*, **29**, 897–907.

7 Butera, R.J., Jr., Wilson, C.G., Delnegro, C.A., and Smith, J.C. (2001) A methodology for achieving high-speed rates for artificial conductance injection in electrically excitable biological cells. *IEEE Trans. Biomed. Eng.*, **48**, 1460–1470.

8 Pinto, R.D., Elson, R.C., Szucs, A., Rabinovich, M.I., Selverston, A.I., and Abarbanel, H.D. (2001) Extended dynamic clamp: controlling up to four neurons using a single desktop computer and interface. *J. Neurosci. Methods.*, **108**, 39–48.

9 Prinz, A.A., Abbot, L.F., and Marder, E. (2004) The dynamic clamp comes of age. *Trends Neurosci.*, **27** (4), 218–224.

10 Strogatz (2001) *Nonlinear Dynamics and Chaos: With Applications to Physics, Biology, Chemistry, and Engineering*, Perseus Books Publishing.

11 Izhikevich (2000) Neural excitability, spiking and bursting. *Int. J. Bifurcat. Chaos*, **10** (6), 1171–1266.

12 Izhikevich (2007) *Dynamical Systems in Neuroscience: The Geometry of Excitability and Bursting*, MIT Press, Cambridge, MA.

13 Khoo, M.C.K. (2000) *Physiological Control Systems Analysis, Simulation, and Estimation*, IEEE Press, Piscataway, NJ.

14 Chen, G., Hill, D.J., and Yu, X. (eds) (2003) *Bifurcation Control Theory and Application, Lecture Notes in Control and Information Sciences*, vol. 293, Springer, New York.

15 Rinzel, J. (1978) On repetitive activity in nerve. *Fed. Proc.*, **44** (15), 2944–2946.

16 Guttman, R., Lewis, S., and Rinzel, J. (1980) Control of repetitive firing in squid axon membrane as a model for a neuroneoscillator. *J. Physiol.*, **305**, 377–395.

17 Paydafar, D., Forger, D.B., and Clay, J.R. (2006) Noisy inputs and the induction of on–off switching behavior in a neuronal pacemaker. *J. Neurophsyiol.*, **96** (6), 3338–3348.

18 Manor, Y. and Nadim, F. (2001) Synaptic depression mediates bistability in neuronal networks with recurrent inhibitory connectivity. *J. Neurosci.*, **21**, 9460–9470.

19 Sorensen, M., Deweerth, S., Cymbalyuk, G., and Calabrese, R.L. (2004) Using a hybrid neural system to reveal regulation of neuronal network activity by an intrinsic current. *J. Neurosci.*, **24** (23), 5427–5438.

20 Olypher, A., Cymbalyuk, G., and Calabrese, R.L. (2006) Hybrid systems analysis of the control of burst duration by low-voltage-activated calcium current in leech heart interneurons. *J. Neurophysiol.*, **96** (6), 2857–2867.

21 Kullmann, P.H., Wheeler, D.W., Beacom, J., and Horn, J.P. (2004) Implementation of a fast 16-bit dynamic clamp using LabVIEW-RT. *J. Neurophysiol.*, **91** (1), 542–554.

22 Milescu, L.S., Yamanishi, T., Ptak, K., Mogri, M.Z., and Smith, J.C. (2008) Real-time kinetic modeling of voltage-gated ion channels using dynamic clamp. *Biophys. J.*, **95** (1), 66–87.

23 Angstadt, J.D. and Calabrese, R.L. (1989) A hyperpolarization-activated inward current in heart interneurons of the medicinal leech. *J. Neurosci.*, **9** (8), 2846–2857.

24 Hill, A.A., Lu, J., Masino, M.A., Olsen, O.H., and Calabrese, R.L. (2001) A model of a segmental oscillator in the leech heartbeat neuronal network. *J. Comput. Neurosci.*, **10**, 281–302.

25 Cymbalyuk, G.S., Gaudry, Q., Masino, M.A., and Calabrese, R.L. (2002) Bursting in leech heart interneurons: cell-autonomous and network-based mechanisms. *J. Neurosci.*, **22** (24), 10580–10592.

26 Hodgkin, A. and Huxley, A. (1952) A quantitative description of membrane current and its application to conductions and excitation in nerve. *J. Physiol.*, **117**, 500–544.

27 Schmidt, J. and Calabrese, R.L. (1992) Evidence that acetylcholine is an inhibitory transmitter of heart interneurons. *J. Exp. Biol.*, **171**, 329–347.

28 Mahowald, M. and Douglas, R. (1991) A silicon neuron. *Nature*, **354** (6354), 515–518.

29 Jung, R., Brauer, E.J., and Abbass, J.J. (2001) Real-time interaction between a neuromorphic electronic circuit and the spinal cord. *IEEE Trans. Neural Syst. Rehabil. Eng.*, **9** (3), 319–326.

30 Rachmuth, G. and Poon, C.S.l. (2008) Transistor analogs of emergent iono-neuonal dynamics. *HFSP J.*, **2** (3), 156–166.

31 Shilnikov, A.S., Calabrese, R.L., and Cymbalyuk, G.S. (2005) Mechanism of bistability: tonic spiking and bursting in a neuron model. *Phys. Rev. E*, **71**, 056214.

32 Szucs, A., Huerta, R., Rabinovich, M.I., and Selverston, A.I. (2009) Robust microcircuit synchronization by inhibitory connections. *Neuron*, **61** (3), 439–453.

33 Kaji, H., Nishizawa, M., and Matsue, T. (2003) Localised chemical stimulation to micropatterned cells using multiple laminar fluid flows. *Lab. Chip.*, **3** (3), 208–211.

34 Peterman, M.C., Mehenti, N.Z., Bilbao, K.V., Lee, C.J., Leng, T., Noolandi, J., Bent,

S.F., Blumenkranz, M.S., and Fishman, H.A. (2003) The Artificial Synapse Chip: a flexible retinal interface based on directed retinal cell growth and neurotransmitter stimulation. *Artif. Organs.*, **27** (11), 975–985.

35 Mourzina, Y., Kaliaguine, D., Schulte, P., and Offenhausser, A. (2006) Patterning chemical stimulation of reconstructed neuronal networks. *Anal. Chim. Acta*, **575** (2), 281–289.

36 Mehenti, N.Z., Fishman, H.A., and Bent, S.F. (2007) A model neural interface based on functional chemical stimulation. *Biomed. Microdevices.*, **9** (4), 579–586.

8
Biomimetic Adaptive Control Algorithms

James J. Abbas

8.1
Introduction

In the quest to design improved tools and devices, engineers are often interested in interpreting patterns of sensory information and/or controlling movements. Not only is the nervous system particularly good at performing these sensory and motor tasks but also the adaptive capabilities of neural tissue endow animals with the ability to learn to improve their performance. For these reasons, there has been great interest in using biomimetic approaches to the engineering design of systems for sensory processing, movement control, and adaptation [1].

8.1.1
Potential to Enhance Capabilities of Engineered Systems

In the engineering world, the primary motivation for pursuing this line of research is the pragmatic goal of overcoming the shortcomings of existing techniques. If existing approaches to sensor processing and control were suitable for the tasks at hand, biomimetic techniques would primarily be a branch of quantitative biology. In many engineering applications, however, there is a growing acknowledgment that existing (nonbiomimetic) techniques have limited capabilities – and those limitations are most apparent in addressing problems that biological systems routinely solve.

In the realm of sensory processing, biological systems perform especially well at tasks such as recognizing faces, comprehending speech in noisy environments, distinguishing odors and tastes, and recognizing shapes and textures – all of which have presented particular challenges to the engineering community. In the realm of control, biological systems display exquisite balance and coordination under a variety of environments, track movements of objects by controlling eye movements within a head and body that are moving rapidly through space, and regulate chemical concentrations in the presence of highly time-varying demands. Meeting these control goals requires integration of information from multiple distributed sources,

Biohybrid Systems: Nerves, Interfaces, and Machines, First Edition. Edited by Ranu Jung.

coordinated activation of several effectors, attention to system dynamics and time delays, and simultaneous consideration of multiple control objectives. The interest in biomimetic control is driven primarily by the need to design systems with these types of capabilities.

Although the sensor processing and motor capabilities of biological systems are attractive to engineering designers, the adaptive capabilities are perhaps most enticing. The ability of biological systems to learn from experience – and in particular to learn in an unsupervised manner – provides them with versatility and autonomy, both of which are of great interest in many engineering fields.

8.1.2 Integrating Engineered Systems with Biological Systems

Technological innovations in the recent past – and on the near horizon – have utilized an increasing degree of interaction with biological systems. This is perhaps most apparent in the field of robotics. While early robots had their impact performing routine and repeatable tasks on the assembly line, many modern robotic systems – in the factory, on the battlefield, and in the operating room – are designed to augment or assist humans in performing tasks.

This interaction with humans (or other biological systems) underlines the need for the engineered system to adapt. In addition to dealing with the uncertainties of the physical environment, the device must also deal with the uncertainties, and variability, that are inherent in biological systems.

The next level of complexity in this progression is the notion of coadaptation – the situation in which both the engineered system and the biological system adapt. The engineered system adapts to account for the specific characteristics of the biological system and the biological system adapts to accommodate the interaction with the robot. In many instances, the role of the engineered system could be to guide the adaptation of the biological system – that is, to take on the role of teacher. The transformation then is from machines that perform tasks, to machines that actively assist humans to perform tasks, to machines that actively train humans to perform tasks.

The best example of the potential role for coadaptation may be in the field of rehabilitation, such as the design of a robot to be used in rehabilitation therapy. In such applications, the variability across a set of individuals is often enhanced by the trauma or neurological disorder. Furthermore, the behavior of the individual is likely to change over time due to spontaneous recovery, the effect of therapy, disease-related degradation, or aging. Therefore, in these applications, the device must have the ability to effectively interact with individuals with a wide range of capabilities and behavior patterns that may change over time.

8.1.3 Focus on the Nervous System

Biological systems use highly integrated control in a number of different realms: genetic, molecular, cellular, biochemical, hormonal, and neuronal. This chapter focuses on neural control systems and how to mimic neural processes that mediate control of complex processes.

8.2 Biomimetic Algorithms

Figure 8.1 presents a general block diagram illustrating basic sensorimotor integration and the role that learning can play in modifying the internal processes to adapt to changing needs. A key feature of this structure is that information derived both from external and from internal structures can be used to modify the sensory processing procedures and/or the control procedures. Although this general structure can describe both nonbiomimetic and biomimetic approaches, the biomimetic approach utilizes biological inspiration in order to implement operations within a given block or to modify or elaborate on the interactions among system components.

Here, we consider four classes of biological models (Figure 8.2) that have potential for use in biomimetic systems: phenomenological input/output models (*input/output models*), models with structural components that are designed to mimic the actions of key regions in the brain and/or spinal cord (*neurostructural models*), models that use simplistic units that represent the most basic properties of neurons (*artificial neural network models*), and models that use detailed biophysical models of neuron compartments, individual neurons, or neuron pools (*biophysical models*). The sections that follow provide a brief description of each of these classes of biological models and indicate some biomedical applications of biomimetic control.

8.2.1 Input/Output Models

Quantitative models of neural control systems date back to the early 1960s when linear systems techniques were used to describe the oculomotor control system [2, 3]

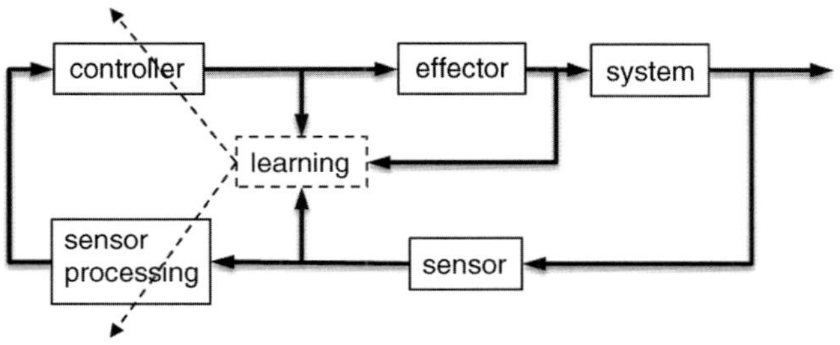

Figure 8.1 General structure for biomimetic adaptive control systems. The feedback pathway (sensor, sensor processing, and control) provides the basic regulatory mechanism that is dependent upon specific parameters for each component. The learning module uses information derived both from sensors and from the effectors and controller output signals; the learning process can alter the operations of the sensor processing block or the controller block.

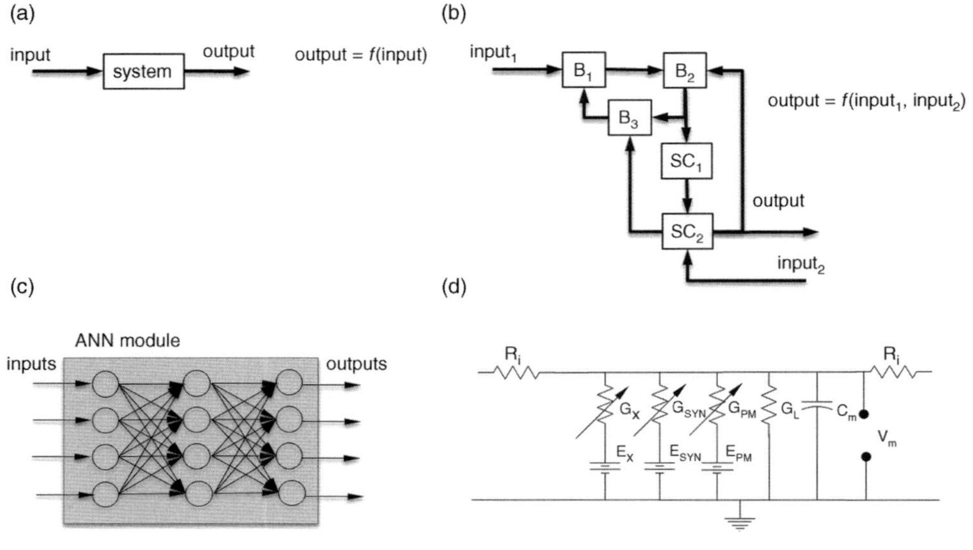

Figure 8.2 (a) Input/output model block diagram structure. In this arrangement, the output is a fixed function of the input; the mapping from input to output may utilize a linear transfer function or may be nonlinear (i.e., a given change in input may not always produce the same change in output). (b) Neurostructural model example. In this hypothetical arrangement, three brain regions (B_1, B_2, B_3) interact with two spinal cord regions (SC_1, SC_2) to process input from two sources (e.g., visual and somatosensory) to determine the output to the effectors, which is delivered through the spinal cord. Several neural models and biomimetic systems utilize such interactions between blocks that represent specific neural centers in the brain or spinal cord. Such blocks have been used to represent small sets of neurons (e.g., column in motor cortex) or large brain regions (e.g., the cerebellum). The equations can take the form of simple linear input/output blocks or high dimensional nonlinear components. (c) Artificial neural network model structure. An ANN module is typically constructed as multiple layers of neurons (circles) that interact via synaptic weights (signals). Weights are adapted using a learning algorithm. In practice, networks may contain tens or hundreds of neurons in a given layer. Connections in the configuration shown are "feedforward" in that the output of a given neuron provides input to the neurons in the following layer. For use in engineering control systems, the ANN module can be used in place of classical transfer functions; for example, the ANN module could be used as the "controller" or "sensory processing" block in Figure 8.1. (d) Basic biophysical model of one compartment of a neuron. This electrical equivalent model of a cell compartment has components that represent the general features of the biophysical model: membrane capacitance (C_m), leak conductance (G_L), ionic conductances (G_X), synaptic conductances (G_{SYN}), and conductances responsible for pacemaker properties (G_{PM}). Each conductance represents a set of channels with a corresponding equilibrium potential (E) that will provide the driving force for current through the channel. The compartment shown, with membrane voltage (V_m), is connected to neighboring compartments through internal resistances (R_i). Additional features are added to represent other channels (neuromodulators, voltage-dependent conductances, calcium-dependent conductances, etc.).

and the cardiovascular control system [4]. Input/output data for the entire system or for each component could be used to determine model parameters for a specified model structure. In this phenomenological approach, the model structure and parameters are selected as those that best fit the input/output data. These models provided mathematically tractable representations of key features of these biological processes and, especially as computing power increased, these models enabled computer simulation studies of the effects of model parameters and interactions between system components. However, these linear representations are clearly limited in their ability to capture many of the critical features of biological control systems.

Biological systems utilize nonlinearities extensively, and in many cases the nonlinear components may be responsible for the versatility, robustness, and stability displayed by biological systems. Incorporating nonlinear components into models with linear dynamic components has in some instances successfully captured the key features of a biological system [5]. To address some of the shortcomings of linear approaches with more generic and widely applicable techniques, some groups have utilized models based on nonlinear kernels [6, 7]. These classes of models and techniques for parameter estimation provide general tools that can be utilized to fit input/output data to nonlinear models.

The primary intent in developing these models was to gain insight into the functioning of the biological control system – that is, they served a basic science objective. These models, however, have been the precursors for many later attempts at mimicking neural control systems. The block diagram structures and sometimes the specific sets of equations have formed the foundation for design of systems to control blood pressure, glucose, arm movements, and locomotion.

8.2.2 Neurostructural Models: Models Based on Regional Neuroanatomy and Neurophysiology

With information derived primarily from investigations of neuroanatomy and lesion studies, the roles of various brain regions in specific tasks have long been understood – at least in a general sense. Movement of our limbs results from interactions among such structures as motor cortex, the cerebellum, the basal ganglia, and spinal motoneuron pools. Inputs from the red nucleus and the vestibular nucleus influence the spinal circuits that regulate posture control through the rubrospinal and vestibulospinal pathways, respectively. These studies have produced block diagram-level models of the brain and spinal cord that describe the roles of various regions and pathways in the control of many physiological functions. In some instances, mathematical representations of the various regions and their interactions have been developed. In contrast to a phenomenological approach, in these models the components and their interactions are established by – and constrained by – the structure, connectivity, and functional interactions identified in neuroanatomical and neurophysiological studies. These types of models have formed the foundation for the development of several biomimetic systems.

Such approaches have been successfully utilized to mimic the oculomotor control system in order to develop systems to track objects that move in the visual field [8, 9]. Several groups have developed models for control of posture [10], pedaling [11, 12], or locomotion [13–17] that use modules to describe the organization in the spinal cord responsible for synergistic activation of muscles.

8.2.3 Artificial Neural Network Models

For the past half century, there has been a broad-based effort to develop computational structures and algorithms that mimic the organization and operations of the nervous system. The research activity has spanned the fields of neuroscience, psychology, computer science, electrical engineering, biomedical engineering, and others.

Many artificial neural network (ANN) systems use only the most basic concept that underlies the organization of the nervous system:

- Individual processing units interact:
  - each neuron receives input from other neurons; some neurons receive inputs from external sources.
  - each neuron output is a (typically nonlinear) function of the weighted sum of the inputs.
  - each neuron sends its output to other neurons.
  - the output of the network is determined by the pattern of activity across a subset of the units.
- Adaptation is produced by changing the strength of interaction of the units via a learning rule.

These basic concepts, which constitute the foundation for the field of artificial neural networks, were first introduced by McCulloch and Pitts in 1943 in the form of the Threshold Logic Unit [17]. Within the field, a wide variety of neuron models (input–output function for each computational unit), network architectures (connectivity pattern among the units), and learning rules (equation that governs adaptation of the weights) have been developed [18]. Although proper selection of neuron model and network architecture is important, the success of artificial neural network techniques can be primarily attributed to the development and clever application of a wide variety of learning algorithms. Some of these algorithms borrow directly from optimization theory, while others borrow directly from biology.

The most commonly used algorithms utilize *gradient descent*, a widely utilized engineering optimization technique that has been incorporated into the neural network structure. These algorithms calculate the gradient of error versus network parameter (weight) and then change the parameter in a direction that reduces error. An important implementation of this approach uses backpropagation, which extends the concept of gradient descent to a multilayer network structure and consequently vastly enhances network capacity. A key aspect of this technique is that it utilizes supervised learning – that is, the algorithm requires knowledge of (or estimates of) the error in the network output. Some of the limitations of this approach include biological plausibility (a problem for purists but not for pragmatists); adaptation can be slow (especially in high dimensional, multilayer implementations), adaptation can be limited in its efficacy (due to the presence of local minima in the error surfaces), and the need to know error in network output.

Another commonly used set of learning paradigms is that of the *Hebbian-type algorithms*. These networks use a basic idea drawn from biological observation (by Donald Hebb): the connection between two neurons is strengthened if both neurons are active at the same time. There are several variations of this idea with many different neuron models and network structures [18]. Some of the limitations of this approach are the need to impose meaning/structure to get effective learning, learning can be slow, and there is a tradeoff between the need to unlearn useless information and the need for persistent excitation.

Extensions of the biological inspiration have used *evolutionary algorithms* to train ANNs. This class of algorithms utilizes a biomimetic approach to parameter selection for the neural network. In this technique, the parameter set is represented as a gene; multiple instances of the parameter set are produced; the fittest instances survive; the next generation of instances are produced using sexual or asexual reproduction, with or without mutations; and the next generation is evaluated. While this technique can be very powerful, one of the limitations is that extensive training times that are required and therefore online implementation may not be practical.

In the biomedical realm, ANN systems have been highly successful at solving some problems that involve pattern classification, adaptive filtering, and, to a lesser degree, adaptive control [19–22].

8.2.4
Biophysical Models: Conductance-Based Models and Beyond

The last century has seen major advances in our understanding of the biophysics of neurons. It is apparent that the functioning of an individual neuron depends upon the capacitive bilipid membrane, passive flow of ionic current through complex dendritic branch systems, voltage-gated ion channels, various neurotransmitter and neuromodulator systems, local ionic concentrations, localized channel distributions, second-messenger systems, and so on. The explosion of data has clearly confirmed the inadequacy of the McCulloch–Pitts formulation as a neuron model. But what are the implications for the design of engineered systems? Can we improve the performance of engineered systems by embedding a higher degree of biological fidelity into the control system?

The increased interest in exploring the utility of neuron models with a higher degree of biological fidelity is driven primarily by two observations: (1) biological systems exhibit capabilities that exceed those of engineered systems that utilize simple neuron models and (2) detailed biophysical models exhibit properties that may be responsible for the impressive capabilities of the biological system.

A simple, single-compartment conductance-based model consists of an RC circuit, with the capacitor representing the bilipid membrane and the resistor representing the pathways for ionic current that are provided by proteins embedded in the membrane. Using this single-compartment conductance-based model as a starting point, several well-documented features could enhance biological fidelity and may endow engineered biomimetic systems with enhanced functionality. Several of these features are listed and described below:

- **Multicompartmental structure:** The single-compartment neuron model characterizes the cell as an isopotential sphere in which the single value for membrane potential represents the state of the neuron at a given time. Biological neurons are complex structures with local variations in electrical potential, ion concentrations, and channel densities. These spatial variations are an integral component of the information processing capabilities of the neuron in that they enable the cell to utilize spatial and temporal information in interpreting synaptic inputs. The effect of an individual postsynaptic potential on the firing pattern of a neuron will depend on the location of the synapse in the cell and on the timing of the event with respect to inputs in neighboring regions of the cell. The multicompartment structure seeks to enhance the information processing capacity of individual neurons by allowing for the spatial variations in electrical potential that then has implications for temporal processing due to the local dynamics [23].
- **Spiking neurons:** Neurons utilize action potentials to transmit information internally (within the neuron), and the arrival of the action potential at a presynaptic terminal is the trigger for the release of neurotransmitter that will then be responsible for generating the postsynaptic potential. Some neuron models represent the firing activity of the neuron as a continuous variable: firing rate. That is, the models assume that information is encoded in variations of the instantaneous firing rate and that the precise timing of individual spikes is inconsequential. In many biological systems and in their biomimetic counterparts, this simplifying assumption may be valid. In some biological systems, however, the precise timing of spikes conveys essential information. Notable examples are in the processing of binaural information for sound source localization [24] and in neuronal adaptation in what has been termed spike timing-dependent plasticity [25]. In these systems, the valuable information that is encoded in the relative timing of spikes may be entirely independent of the information that is encoded in firing rate. Engineered systems that utilize spiking neurons may be able to effectively utilize this additional dimension of information capacity to enhance the processing power of biomimetic systems.
- **Synaptic dynamics:** In many artificial neural network systems and neural models, synaptic interactions are modeled as instantaneous processes. In models that do not use spiking neurons, postsynaptic effects are often determined entirely by the instantaneous firing rate of the presynaptic neuron. In models that use spiking neurons, the postsynaptic potential that results from a presynaptic spike is often incorporated as a dynamic process. The magnitude and time profile of the postsynaptic potential can depend upon receptor dynamics, second-messenger

dynamics, and dendritic spine shape. One consequence of the time-dependent response of the synapse is that it can have a strong influence on the spatiotemporal integration of synaptic inputs in the cell. Perhaps more importantly, the processes that underlie the synaptic dynamics may provide a powerful mechanism for nonlinear information processing and integration. The general theme is prevalent in biological systems – complexity in a pathway provides opportunities for interaction and complex processing. Biomimetic systems may be able to utilize these processes to enhance the richness and subtleties of information processing and control.

- **Synaptic reversal potentials:** In artificial neural networks and many models of synaptic interaction, synapses are defined as excitatory or inhibitory by the sign of the synaptic weight. Furthermore, the postsynaptic effect depends only on presynaptic activity and the strength of the synapse. In biological systems, however, synaptic effect is strongly influenced by the state of the postsynaptic neuron. Synaptic inputs cause channels to open; current flow through the channel depends upon the equilibrium potential for the channel and the local membrane potential. In this manner, both the direction and the magnitude of synaptic currents are affected by the postsynaptic state. By utilizing such mechanisms for state-dependent responses, biomimetic control systems may be able to react more appropriately to multidimensional inputs.
- **Neuromodulators and multiple transmitter systems:** An individual neuron utilizes a wide array of protein channels that regulate ionic flow; many of these are responsive to the chemical environment of the cell. Some of these chemicals are released from other neurons and exert their effect through typical synaptic structures; on its postsynaptic side, an individual neuron may use a variety of transmitter systems across its set of synapses. Other chemicals exert their effect in a broad and more diffuse manner and have what has been termed a neuromodulatory effect. Some of these neuromodulators are released from other neurons or their support cells, while others are hormones released by the endocrine system. Typically, neuromodulators exert their effects with dynamics that are much slower than neurotransmitters. This neuromodulatory mechanism enables individual neurons to process synaptic interaction in a highly state-dependent manner and provides yet another mechanism for neuronal integration across systems or subsystems. In biomimetic systems, there is a strong interest in incorporating mechanisms for state-dependent responses since this is a hallmark of biological systems and one that is critical for performing complex functions; utilizing models of neuromodulatory systems may provide such capabilities.
- **Intracellular calcium pools:** Local calcium concentrations play a large role in regulating the dynamics of intracellular pathways and their interactions [23]. Modeling these effects has demonstrated that intracellular variations in calcium concentrations can influence dendritic integration and firing properties. In biomimetic systems, this aspect of cellular function may be able to enhance the complexity and capabilities of individual neurons and their interactions.
- **Presynaptic inhibition:** The impact of synaptic location is perhaps nowhere more apparent than in the arrangement that allows presynaptic inhibition. In this

situation, one synaptic input can selectively silence the action of another. The anatomical relationship is the primary factor in determining this interaction; the dynamics of the individual postsynaptic potentials can influence the degree and timing of the effect. From a control systems perspective, this mechanism gives a highly selective means of implementing and dynamically modifying the integration of information from several sources. It is apparent that nature has utilized this mechanism and anatomical arrangement extensively; in biomimetic systems, this structural arrangement may provide an efficient and flexible building block for systems that exhibit rich sensorimotor capabilities.

Collectively, these and other features of biological neurons are, at least in part, responsible for some of the capabilities of biological systems that are of interest to biomimetic engineers [24, 26]. Basic models for these features have been developed; a key challenge is to incorporate these features into sensory processing and control structures so that they can be utilized effectively in biomimetic systems. More information regarding these models is presented in Chapter 2.

8.2.5
Central Pattern Generators

For the control of cyclic tasks such as locomotion, many biological systems use circuits that have been described as central pattern generators (CPGs) [27–29]. These circuits consist of a group of neurons that interact to produce an oscillatory pattern of activity. Although in the purest sense of the term, a CPG is capable of generating oscillatory activity in the absence of sensory input, this arrangement is never (or at least rarely) observed *in vivo* in biological systems and may be of limited use in biomimetic systems. Inputs to the CPG in the form of afferent feedback can couple the neuronal dynamics to that of the plant (system) and provide information to modify CPG signals in order to regulate outputs of the plant.

Investigations and modeling of CPG circuitry of several organisms have provided insight into the mechanisms used by CPGs and suggest potential advantages for using CPG circuits in biomimetic systems. Some of these characteristics are properties of cells (examples are provided in the set of equations given in Table 8.1); others are properties of the circuit (examples are provided in the neural circuit diagram given in Figure 8.3):

- **Plateau potentials:** Several cells have the intrinsic capacity to generate a sustained response to a transient input. Specialized membrane proteins provide this mechanism, which can be utilized effectively in CPG circuits to generate oscillatory patterns on timescales that are several orders of magnitude greater than that of an action potential or membrane time constant. During the depolarizing phase of these oscillations, bursts of action potentials produce inputs to other cells in the circuit and the internal mechanics of generating action potentials influence the intrinsic membrane oscillations and the burst properties.
- **Voltage-dependent channels:** Many ionic channels exhibit a voltage-dependent response that, similar to the effect of the postsynaptic potential, can provide a

Table 8.1 Example CPG Neuron equations.

Membrane dynamics:

$$\mathrm{d}V_{\mathrm{m}_i}(t)/\mathrm{d}t = (I_{\mathrm{syn}_i}(V_{\mathrm{m}}, t) + I_{\mathrm{PM}_i}(V_{\mathrm{m}}, t) + I_{\mathrm{inj}_i} - V_{\mathrm{m}_i}(t)/R_{m_i})/C_{m_i} \quad (8.1)$$

Output function (firing rate):

$$y_i(t) = \frac{1}{1 + \mathrm{e}^{-2m_i(V_{\mathrm{m}_i}(t) - V_{\mathrm{o}_i})}} \quad (8.2)$$

Membrane currents:

$$I_{\mathrm{syn}_i}(t) = \sum y_j(t) g_{\mathrm{syn}_{ij}} (E_{\mathrm{syn}_{ij}} - V_{\mathrm{m}_i}(t)) \quad (8.3)$$

$$I_{\mathrm{PM}_i}(t) = I_{\mathrm{NMDA}_i}(t) + I_{\mathrm{KCa}_i}(t) \quad (8.4)$$

Burst initiation:

$$I_{\mathrm{NMDA}_i}(t) = K_{\mathrm{NMDA}_i} g_{\mathrm{NMDA}_i} p_i(t)(E_{\mathrm{NMDA}_i} - V_{\mathrm{m}_i}(t)) \quad (8.5)$$

$$\frac{\mathrm{d}p_i(t)}{\mathrm{d}t} = A_{\alpha i}\mathrm{e}^{((V_{\mathrm{m}_i} - E_r)/C_{\alpha_i})}(1 - p_i(t)) - A_{\beta_i}\mathrm{e}^{((E_r - V_{\mathrm{m}_i})/C_{\beta_i})} p_i(t) \quad (8.6)$$

Burst termination:

$$I_{\mathrm{KCa}_i}(t) = g_{\mathrm{KCa}_i}[Ca_{\mathrm{NMDA}_i}(t)](E_{\mathrm{KCa}_i} - V_{\mathrm{m}_i}(t)) \quad (8.7)$$

$$\frac{\mathrm{d}[\mathrm{Ca}_{\mathrm{NMDA}}(t)]_i}{\mathrm{d}t} = p_i(t)\varrho_{\mathrm{NMDA}_i} K_{\mathrm{NMDA}_i}(E_{\mathrm{CaNMDA}_i} - V_{\mathrm{m}_i}(t)) - \delta_{\mathrm{NMDA}_i}[Ca_{\mathrm{NMDA}}(t)]_i \quad (8.8)$$

This set of equations provides a model neuron with intrinsic oscillatory capabilities. When combined with a network interaction such as the one depicted in Figure 8.1, biological systems utilize a layered structure that uses interactions between intrinsic cellular properties and network properties to generate stable, yet flexible and adaptable oscillatory patterns. This model includes a capacitive membrane with multiple channels for current to flow (Equation 8.1); a neuron output that describes firing rate as an instantaneous function of membrane potential (Equation 8.2). The membrane currents include synaptic currents that utilize synaptic reversal potentials (Equation 8.3) and pacemaker currents that provide the intrinsic oscillatory capacity. In this model, the burst initiation is provided by NMDA channels (Equations 8.5 and 8.6) and the burst termination is provided by the calcium-dependent potassium channels (Equations 8.7 and 8.8). Collectively, this set of equations endows the cell with oscillatory capabilities. Such mechanisms are exhibited in biological oscillators and have been utilized both in biological models and in some biomimetic control systems.

mechanism for the cell to process inputs in a state-dependent manner. Such channels have been used in CPG models [30, 31] and have been used in biomimetic control system design to contribute to the pattern-generating capabilities of the neural circuit [32].

- **Reciprocal inhibition:** In cells that utilize intrinsic or network mechanisms to produce sustained plateau potentials, there must be some mechanism (or mechanisms) in place to shut down the burst. Voltage-dependent channels with

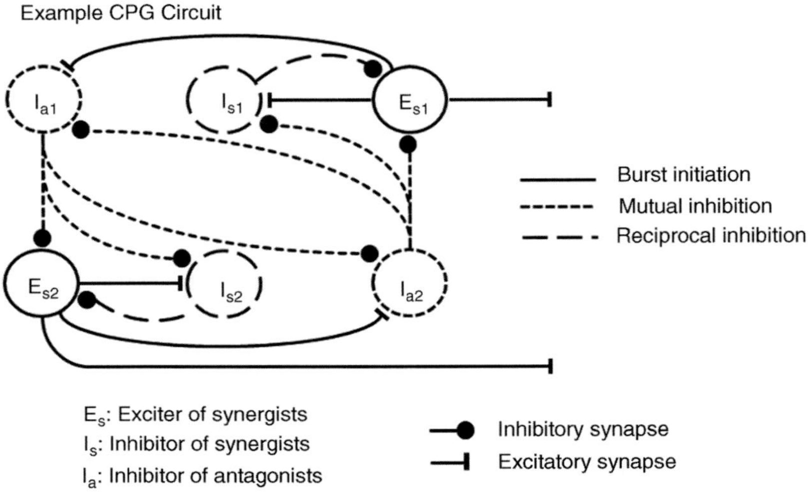

Figure 8.3 Example CPG circuit that demonstrates some of the fundamental network mechanisms used by CPGs to produce stable, yet versatile oscillatory patterns. One key component is to have a mechanism for burst initiation. In this circuit, this is provided by the Es neurons, which have intrinsic properties that tend to initiate a burst (see example equations in Table 8.1). The second component is a mechanism to alternate between sides of the oscillator (e.g., flexor/extensor or left/right oscillations in circuits for locomotion). Biological systems commonly utilize a mutually inhibitory arrangement in which opposite sides of the oscillator inhibit each other, as shown in this example. A third component is a mechanism for terminating a burst. In this example, a reciprocally inhibitory network arrangement provides this feature. On each side of the oscillator, as the Es neuron fires, it excites the Is neuron, which in turn inhibits the Es neuron. These mechanisms are commonly exhibited in biological oscillators and have been utilized both in biological models and in some biomimetic control systems.

time-dependent responses can provide an intrinsic mechanism, but this is often supplemented by a network mechanism of reciprocal inhibition. In this arrangement, one neuron excites another, which in turn inhibits the first. The membrane time constant and other dynamic processes involved in cellular interactions and responses result in a delay between the initial activity and the recurrent inhibition. In this manner, the network arrangement can self-regulate activity in a manner that affects overall oscillation period as well as the activity patterns of individual cells in the circuit.

- **Mutual inhibition:** At the small circuit level, biological systems utilize the mechanism of mutual inhibition, in which two cells (or two groups of cells) inhibit each other. When combined with internal dynamics and/or time delays, the mutual inhibitory structure can form the foundation of an oscillatory circuit. This mechanism has been demonstrated in many models of neural circuits and has been used in biomimetic control system design [22, 32–43].

These cellular and network properties endow the CPG with several capabilities that may be highly desirable for an engineered biomimetic system: stable oscillations, frequency modulation, phase-dependent sensory integration, and phase resetting [44–47]. Each of these has been demonstrated in mathematical models of CPGs and has been utilized in biomimetic systems [32–43].

8.3 Discussion

In the implementation of any adaptive system, there are risks of adapting in an inappropriate manner. One key issue revolves around the decision of when to adapt and how strongly to adapt. To operate autonomously, the adaptive system must determine when the information derived from sensors is not only reliable but also useful. Context can play a critical role in determining the most appropriate way to adapt. For example, information derived from sensors on the feet just after the foot hits the ground may be very useful in adapting a specific set of parameters of the CPG circuit whereas the signal from the same set of sensors may not be useful in other phases of the gait cycle. Information derived from sensors with a high noise level may lead to adaptation that degrades overall performance. Biomimetic approaches to adaptive weighting of sensory inputs for both real-time control and real-time adaptation may provide advantages over existing techniques. In particular, biomimetic approaches may enable integration of multiple sensors and multiple sensory modalities in order to fine-tune the adaptive processes.

As we design systems that are intended to coadapt, this adds a new dimension to the complexity of the adaptive process and raises important questions about the stability of the adaptive processes. In psychophysical studies of humans interacting with machines, the general observation is that people learn best how to use a tool if that tool behaves reliably and in a repeatable manner, thus suggesting that adaptation in the device could be detrimental to overall performance. As we design adaptive biomimetic systems, how will the adaptability that is engineered into the biomimetic device affect the user's ability to use the device? How will the adaptability affect the user's learning process? Similar concerns are raised about the reciprocal process – that is, how will the changes in performance due to biological learning affect the adaptive procedures engineered into the biomimetic system? In this coadaptive environment, there is great potential for instabilities in the adaptive processes. It may be possible to address these concerns by simply avoiding rapid adaptation; if the time constants on the learning processes are slow enough, instabilities may be avoided. However, effective coadaptation may require new theoretical developments to maximize learning while ensuring stability.

For many applications that involve interactions with people, the complexity and potential pitfalls of implementing adaptive engineered systems may contraindicate (or at least call into question) their use. A valid approach may be to design a nonadaptive engineered system and let the biological system handle all adaptation. In the design of the next generation of "smart" prostheses, the question has been raised: should we engineer the smarts into the prosthesis or should we design a highly capable "dumb" (nonadaptive) prosthesis and let the brain learn how to best use it [48]?

8.4
Future Developments

It is apparent that many biological systems exhibit capabilities that far exceed the best that today's engineered systems have to offer. Although there is growing interest in utilizing biomimetic approaches to develop the next generation of high-performance engineered systems, it is not clear how best to proceed. As we move along the spectrum from today's engineering devices to high-fidelity biomimetic systems, there is an important tradeoff between the potential benefits of biological fidelity and the costs of the complexity of design and implementation.

The cost/benefit tradeoff is likely to be greatly improved by developments in the following key areas:

- The development of theoretical frameworks and mathematical tools can provide mathematical/engineering rigor to complex biomimetic systems. The growing interest and activity in the fields of computational neuroscience and neural modeling (Chapter 2) will provide new algorithms to implement high-fidelity models of individual neurons and their interactions.
- The development of neuromorphic hardware (e.g. Chapter 3 and Chapter 9) [34, 41] can facilitate implementation of biomimetic systems by providing the capability of real-time simulation of complex biomimetic models. Such processing capability is not only critical for real-time control but it will also accelerate the development of the knowledge base for biomimetic systems.
- The development of high-density/high-fidelity neural interfaces (e.g. Chapter 6 and Chapter 7) will enable the integration of biomimetic and biological systems. In designing machines that interact with people, neural interfaces provide a means to directly connect the engineered system to the nervous system. If used appropriately, these interfaces may provide a high degree of integration between the user and the machine [48, 49]. A high-quality interface may not only improve the performance of the biohybrid systems but may also make them easier to learn to use. The development of these interfaces will provide new opportunities for the biomimetic community and it will provide a new set of challenges.
- The development of an improved understanding of neuroplasticity will enhance our ability to mimic it and promote it. Although there has been great progress in the development of the knowledge base that describes learning in the nervous system, there are still several critical gaps. A more detailed and nuanced understanding of the multiple mechanisms involved in learning can be incorporated into the next generation of adaptive algorithms. Perhaps more importantly, that knowledge can also be used to design the biomimetic system such that its interactions with the biological system can maximize learning in the biological system.

References

1 Bar-Cohen, Y. (2006) Biomimetics: using nature to inspire human innovation. *Bioinsp. Biomim.*, **1** (1), P1–P12.

2 Young, L.R. and Stark, L. (1963) A discrete model for eye tracking movements. *IEEE Trans. Mil. Electron.*, **7** (2), 113–115.

3 Robinson, D.A. (1973) Models of the saccadic eye movement control system. *Biol. Cybern.*, **14** (2), 71–83.

4 Katona, P.G., Barnett, G.O., Terry, B.S., Katona, P.G., Poitras, J.W., Barnett, G.O., and Terry, B.S. (1970) Cardiac vagal efferent activity and heart period in the carotid sinus reflex. *Am. J. Physiol.*, **218** (4), 115–133.

5 Cook, G. and Stark, L. (1967) Derivation of a model for the human eye-positioning mechanism. *Bull. Math. Biol.*, **29** (1), 153–174.

6 Watanabe, A. and Stark, L. (1975) Kernel method for nonlinear analysis: identification of a biological control system. *Math. Biosci.*, **27** (1–2), 99–108.

7 Marmarelis, V.Z. (1997) Modeling methodology for nonlinear physiological systems. *Ann. Biomed. Eng.*, **25** (2), 239–251.

8 Lenz, A. *et al.* (2008) An adaptive gaze stabilization controller inspired by the vestibulo-ocular reflex. *Bioinsp. Biomim.*, **3** (3), 035001.

9 Shibata, T. and Schaal, S. (2001) Biomimetic gaze stabilization based on feedback-error-learning with nonparametric regression networks. *Neural Netw.*, **14** (2), 201–216.

10 Ting, L.H. and Macpherson, J.M. (2005) A limited set of muscle synergies for force control during a postural task. *J. Neurophysiol.*, **93** (1), 609–613.

11 Raasch, C.C. *et al.* (1997) Muscle coordination of maximum-speed pedaling. *J. Biomech.*, **30** (6), 595–602.

12 Raasch, C.C. and Zajac, F.E. (1999) Locomotor strategy for pedaling: muscle groups and biomechanical functions. *J. Neurophysiol.*, **82** (2), 515–525.

13 Ivanenko, Y.P., Poppele, R.E., and Lacquaniti, F. (2004) Five basic muscle activation patterns account for muscle activity during human locomotion. *J. Physiolo.*, **556** (1), 267–282.

14 Ivanenko, Y.P., Poppele, R.E., and Lacquaniti, F. (2009) Distributed neural networks for controlling human locomotion: lessons from normal and SCI subjects. *Brain Res. Bull.*, **78** (1), 13–21.

15 Ivanenko, Y.P. *et al.* (2007) Modular control of limb movements during human locomotion. *J. Neurosci.*, **27** (41), 11149–11161.

16 Neptune, R.R., Clark, D.J., and Kautz, S.A. (2009) Modular control of human walking: a simulation study. *J. Biomech.*, **42** (9), 1282–1287.

17 McCulloch, W. and Pitts, W. (1943) A logical calculus of the ideas immanent in nervous activity. *Bull. Math. Biol.*, **5** (4), 115–133.

18 Hagan, M.T., Demuth, H.B., and Beale, M.H. (1996) *Neural Network Design*, PWS Publisher.

19 Narendra, K.S. and Parthasarathy, K. (1990) Identification and control of dynamical systems using neural networks. *IEEE Trans. Neural Netw.*, **1** (1), 4–27.

20 Riess, J. and Abbas, J.J. (2000) Adaptive neural network control of cyclic movements using functional neuromuscular stimulation. *IEEE Trans. Rehabil. Eng.*, **8** (1), 42–52.

21 Davoodi, R. and Andrews, B.J. (1999) Optimal control of FES-assisted standing up in paraplegia using genetic algorithms. *Med. Eng. Phys.*, **21** (9), 609–617.

22 Prentice, S.D., Patla, A.E., and Stacey, D.A. (1998) Simple artificial neural network models can generate basic muscle activity patterns for human locomotion at different speeds. *Exp. Brain Res.*, **123** (4), 474–480.

23 Siegel, M., Marder, E., and Abbott, L.F. (1994) Activity-dependent current distributions in model neurons.

Proc. Natl. Acad. Sci. U S A, **91** (24), 11308–11312.

24 Furukawa, S. and Middlebrooks, J.C. (2002) Cortical representation of auditory space: information-bearing features of spike patterns. *J. Neurophysiol.*, **87** (4), 1749–1762.

25 Dan, Y. and Poo, M.-M. (2006) Spike timing-dependent plasticity: from synapse to perception. *Physiol. Rev.*, **86** (3), 1033–1048.

26 Capaday, C., Cody, F.W.J., and Stein, R.B. (1990) Reciprocal inhibition of soleus motor output in humans during walking and voluntary tonic activity. *J. Neurophys.*, **64**, 607–616.

27 Grillner, S. and Dubuc, R. (1988) Control of locomotion in vertebrates: spinal and supraspinal mechanisms. *Adv. Neurol.*, **47**, 425–453.

28 Grillner, S. and Wallen, P. (1985) Central pattern generators for locomotion, with special reference to vertebrates. *Annu. Rev. Neurosci.*, **8**, 233–261.

29 Goulding, M. (2009) Circuits controlling vertebrate locomotion: moving in a new direction. *Nat. Rev. Neurosci.*, **10** (7), 507–518.

30 Brodin, L. *et al.* (1991) Computer simulations of *N*-methl-D-aspartate receptor-induced membrane properties in a neuron model. *J. Neurophys.*, **66** (2), 473–484.

31 Ekeberg, O. (1993) A combined neuronal and mechanical model of fish swimming. *Biol. Cybern.*, **69**, 363–374.

32 Abbas, J.J. (1996) Using neural models in the design of a movement control system, in *Computational Neuroscience* (ed. J.M. Bower) Academic Press, New York, pp. 305–310.

33 Abbas, J.J. and Chizeck, H.J. (1995) Neural network control of functional neuromuscular stimulation systems: computer simulation studies. *IEEE Trans. Biomed. Eng.*, **42** (11), 1117–1127.

34 Vogelstein, R.J. *et al.* (2008) A silicon central pattern generator controls locomotion *in vivo*. *IEEE Trans. Biomed. Circuits Syst.*, **2** (3), 212–222.

35 Vogelstein, R.J. *et al.* (2006) Dynamic control of the central pattern generator for locomotion. *Biol. Cybern.*, **95** (6), 555–566.

36 Taga, G. (1995) A model of the neuro-musculo-skeletal system for human locomotion. II Real-time adaptability under various constraints. *Biol. Cybern.*, **73** (2), 113–121.

37 Taga, G. (1995) A model of the neuro-musculo-skeletal system for human locomotion I: emergence of basic gait. *Biol. Cybern.*, **73**, 97–111.

38 Ijspeert, A.J. and Kodjabachian, J. (1999) Evolution and development of a central pattern generator for the swimming of a lamprey. *Artif. Life*, **5** (3), 247–269.

39 Ayers, J. and Witting, J. (2007) Biomimetic approaches to the control of underwater walking machines. *Philos. Trans. R. Soc. Lond. A*, **365** (1850), 273–295.

40 Ijspeert, A.J. *et al.* (2007) From swimming to walking with a salamander robot driven by a spinal cord model. *Science*, **315** (5817), 1416–1420.

41 Jung, R., Brauer, E.J., and Abbas, J.J. (2001) Real-time interaction between a neuromorphic electronic circuit and the spinal cord. *IEEE Trans. Neural Syst. Rehab. Eng.*, **9** (3), 319–326.

42 Verdaasdonk, B.W., Koopman, H.F.J.M., and Helm, F.C.T. (2009) Energy efficient walking with central pattern generators: from passive dynamic walking to biologically inspired control. *Biol. Cybern.*, **101** (1), 49–61.

43 Verdaasdonk, B., Koopman, H., and Helm, F. (2006) Energy efficient and robust rhythmic limb movement by central pattern generators. *Neural Netw.*, **19** (4), 388–400.

44 Ting, L.H. *et al.* (1998) Sensorimotor state of the contralateral leg affects ipsilateral muscle coordination of pedaling. *J. Neurophysiol.*, **80** (3), 1341–1351.

45 Capaday, C. and Stein, R.B. (1986) Amplitude modulation of the soleus H-reflex in the human during walking and

standing. *J. Neurosci.*, **6** (5), 1308–1313.

46 Zehr, E.P., Hesketh, K.L., and Chua, R. (2001) Differential regulation of cutaneous and H-reflexes during leg cycling in humans. *J. Neurophysiol.*, **85** (3), 1178–1184.

47 Hultborn, H. and Nielsen, J.B. (2007) Spinal control of locomotion: from cat to man. *Acta Physiol.*, **189** (2), 111–121.

48 Peckham, P.H. (2007) *Smart Prosthetics: Exploring Assistive Devices for the Body and Mind*, National Academies Press.

49 Fagg, A.H. *et al.* (2007) Biomimetic brain machine interfaces for the control of movement. *J. Neurosci.*, **27** (44), 11842–11846.

9
Neuromorphic Hardware for Control

Kevin Mazurek, R. Jacob Vogelstein, and Ralph Etienne-Cummings

Neuromorphic hardware emulates functions of the nervous system in silicon integrated circuits in an attempt to replicate the performance of its biological counterpart. The term neuromorphic was coined by Carver Mead in the early 1990s and is commonly associated with low-power analog circuits that operate in an analogous fashion to biological systems [1]. This hardware is commonly found in biomimetic systems emulating processes such as vision, audition, locomotion, and elementary processes such as neural encoding, discrimination, and adaptation [2–11]. This emulation of lower-level processes can then be used to provide functional control of an internal or external system. This is advantageous in biomimetic systems because it ensures that computations are performed as would be observed in nature. Information in Refs [1, 12, 13] and Chapter 3 provide a more detailed background about the fundamental design and analysis of these building blocks. This chapter provides detailed examples of systems where neuromorphic hardware is used for functional control. In these applications, the control signal is one that another system uses to determine its next course of action. This may involve an analog signal from a neuromorphic vision system that tracks an object [4] or a spiking signal from external sensors for controlling the stimulation waveform of a neuroprosthesis [10]. Mobile applications typically utilize neuromorphic circuits because they operate effectively while consuming low power that is ideal when attempting to conserve battery power. The applications discussed in the following sections include locomotion, audition, and vision. These are a sample of the many applications in existence that incorporate neuromorphic hardware in their biomimetic systems.

9.1
Neuromorphic Hardware for Locomotion

When designing and using neuromorphic hardware, the application for which it is being used determines the type of biological components to emulate and incorporate in the device. For applications involving locomotion, neural circuitry that controls

Biohybrid Systems: Nerves, Interfaces, and Machines, First Edition. Edited by Ranu Jung.

sequential activation of muscles is the key biological process to emulate. Neural circuitry that can produce periodic outputs with periodic forcing forms central pattern generators (CPGs). CPGs for locomotion produce efferent neural signals to control body appendages for movement observed during walking, running, or any other type of gait [14–20]. Their activity can also be modified by sensory afferents. The afferent sensory feedback aids the CPG in fine-tuning the legged movements to ensure robustness against perturbations such as slipping or tripping.

The use of neuromorphic hardware for locomotor control has often been used in the realm of robotics. There is a need to provide control mechanisms to determine how to activate different legs of a robot, let it be bipedal, quadrupedal, and so on. Digital processors are used to implement algorithms that coordinate the activation of different actuators or DC motors (depending on what operates the robotic limbs). For neuromorphic systems, it is desirable to replicate the control mechanisms that aid organisms in biology, the CPG. In doing so, the movements attained by the robot will appear more biologically realistic, which is the purpose of using neuromorphic hardware. By achieving this biofidelity, it allows for the robot to respond to different environmental changes similar to those observed in nature.

There are several neuromorphic devices that emulate the CPG of different animals in order to control different legged robots. Such devices provide control to bipedal, quadrupedal, and hexapedal robots to aid in locomotion [9, 21, 22]. The example of neuromorphic hardware emulating bipedal movements uses a programmable network of silicon neurons to generate output-spiking patterns similar to what is observed in biological recordings [23]. In order to generate outputs to accurately control a bipedal robot, the inhibitory and excitatory weights between neurons are set appropriately to achieve proper timing and sensory feedback from external signals aided in the activation/deactivation of different neurons. In this system, the parameter space is quite large (several synaptic weights need to be set to achieve the desired output). As a result, genetic algorithms are implemented to aid the programming of the SiCPG to obtain walking patterns at certain stepping frequencies faster than a "guess and check" method [24]. A more detailed explanation of the SiCPG will be provided later in this section.

For quadrupedal CPGs, the neuromorphic hardware generates output control using feedforward techniques [22]. The different gait patterns generated are controlled by setting certain input values. The device consists of four coupled oscillators that control the frequency, pulse width, and oscillatory delay through appropriate settings of input voltage values. The oscillators intend to mimic the motor unit outputs observed in biology. These oscillator circuits require bias voltages to determine the overall stepping frequency, stepping pattern, and direction of gait generated. The output patterns from this neuromorphic chip can control DC motors of a four-legged robot to recreate different types of gaits [22]. The difference between the generated gaits is the phasic relationship between the four moving legs. For example, the order in which the four legs move during walking is different from that during a cantor. The device uses a support vector machine (SVM) to learn the appropriate values for each type of gait, as an alternative to optimization techniques such as genetic algorithms previously mentioned.

Hexapedal walking is achieved through neuromorphic hardware by using cellular neural networks (CNNs) to determine the pattern of output activation [9]. These CNNs are implemented in VLSI to generate the desired outputs similar to the biological CPG of a hexapod. (More information about CNNs and the theory behind them can be found in Ref. [25].) The hexapedal device generates different gait patterns based on reconfigurable input variables. The outputs of the device control a hexapedal robot to realize the desired gait patterns and movements.

The previous applications each incorporate a neuromorphic CPG chip to control the movements of a robot. Another way to use this technology is to create a controller that acts as a neuroprosthesis for restoring locomotion in living systems [10]. One such application using the same SiCPG chip as mentioned before is described in the following sections.

9.1.1 A Biohybrid System for Restoring Quadrupedal Locomotion

Spinal cord injury (SCI) occurs in approximately 12 000 individuals in the United States each year [26]. When the injury occurs above the lumbosacral cord, there is loss of activation of the spinal motor pattern generating circuitry in the lumbosacral cord by descending input from the brain. This loss of activation results in paralysis. However, the neural circuitry controlling locomotor movements remains intact as long as the lesion of paralysis occurs above the lumbosacral region of the spinal cord [14]. By replacing the lost brain control of this circuitry with an external device, a biohybrid system can be built to restore lost locomotor control.

9.1.2 Silicon Neural Network Design

The silicon CPG chip (described earlier for controlling a bipedal robot) for restoring locomotion consists of a membrane capacitance, interconnecting synapses, and an axon hillock [23]. A block diagram of these components in an individual spiking neuron is shown in Figure 9.1. There are four different types of synapses available on the chip allowing for different means of neural communication. Four analog and four digital synapses are present that allow for external inputs to excite or inhibit certain neurons. There are also 10 feedback or recurrent synapses to interconnect each neuron with one another and with itself. An additional test synapse is available for each neuron but is typically used for testing purposes. Each feedback synapse has a programmable digital weight that controls the strength of excitation or inhibition each neuron imposes. The digital weights are used to program a current mode analog to digital converter [27], which adds charge to a neuron's membrane voltage. Figure 9.2a depicts the current steering circuitry that allows for this programmability. The synaptic value is an 8-bit word (represented as {S0, ..., S7}) and passes scaled versions of the reference current to the neuron's membrane voltage node. An additional bit is used to specify whether the programmed synaptic weight is excitatory or inhibitory. This determines the direction the current moves toward the

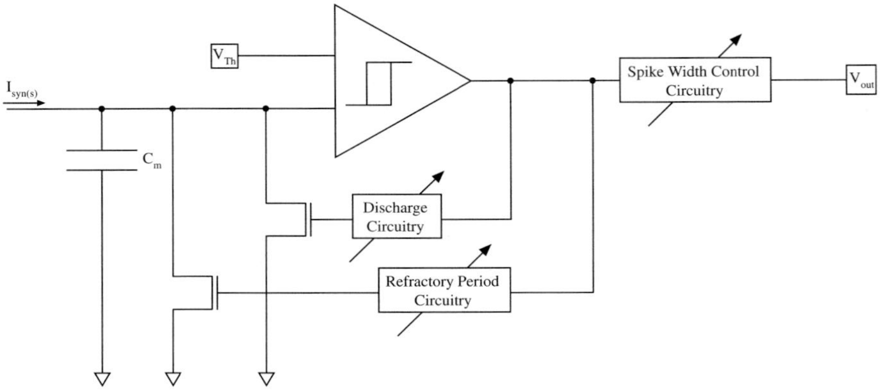

Figure 9.1 Block diagram of the silicon neurons in the SiCPG chip. (Adapted from Ref. [10].)

membrane potential, that is, excitatory adds charge while inhibitory removes it. A hysteretic comparator is implemented to represent the axon hillock and transitions between high and low states depending on the membrane voltage on the input capacitor.

There are additional circuit components representing a refractory period, discharge rate, and spike width control that can be programmed to fine-tune the shape of the spike. The different parameters available to modify the spiking properties of each neuron provide the ability to output different rhythmic spiking patterns similar to those observed in biological systems such as the locomotor CPG.

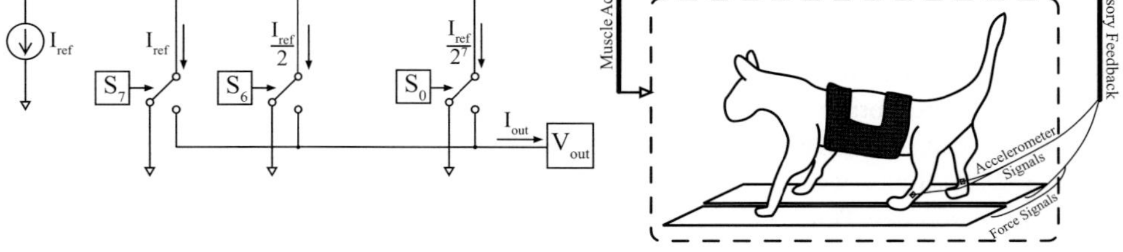

Figure 9.2 (a) Circuit diagram for the programmable synapse. The output current is passed to the capacitor to either increase or decrease its voltage potential (depending on whether the synapse is excitatory or inhibitory). (b) The preprogrammed CPG controller determines the muscles to be stimulated at a given time (*top*). The sensory feedback from the accelorometers and force plates the cat is walking on triggers the transition from stance-to-swing for each leg (*bottom*). (Adapted from Ref. [10].)

9.1.3
Using the Chip for Locomotor Control

By appropriately setting the programmable synaptic weights of the different silicon neurons, it is possible to produce rhythmic spiking outputs representing different muscle activation during locomotion. The transitions between active neurons depend on the synaptic weights that are set both with digital recurrent synapses and with external sensory (analog or digital) synapses. The validity of this device is tested on anesthetized cats to ensure its appropriate performance, and such an experiment is described as follows [10]. The male cats are anesthetized through inhalation of isoflurane, and an intravenous catheter is placed for the administration of somnotol, a barbiturate anesthetic. The SiCPG controls the stimulation pattern of intramuscular stimulation (IMS) electrodes in order to create the desired walking movements. The IMS electrodes are inserted near the six major motor points of the hindlimbs to be able to generate sufficient walking movements. Movements are considered sufficient if they are capable of producing an appropriate amount of weight bearing force and the necessary motion to move the body forward in an appropriate walking pattern.

The cat is placed on a sling attached to a trolley where it can walk along a walkway. The walkway is attached to two parallel force plates that record the ground reaction forces (GRFs) of each leg, while accelerometers are placed on the hind limbs of each leg. These two signals provide feedback information to determine when each leg should make a stance-to-swing or swing-to-stance transition. Proper filtering and calibrating of threshold values enables the SiCPG to make these swing and stance transitions during the *in vivo* experiments. The SiCPG will transition to a swing state or a stance state based on predetermined threshold values measured prior to implementing overground walking. The transitions occur when the recorded sensory signals cross these thresholds, which allows for a change in which leg is in stance or swing, thus producing the appropriate walking pattern of alternating extension and flexion between each leg.

As demonstrated in Ref. [10], a total of 57 trials were conducted on the three cats to determine the efficacy of the controller in producing organized and stable walking (see Figure 9.3). A degree of variability between experiments and between cats relates to muscle fatigue, electrode placement, walkway friction, and sensory thresholds. However, successful walks were exhibited in certain trials when the SiCPG was programmed appropriately. Implementing a more complex CPG model to accommodate for the variability of the environment (such as perturbations and walkway resistance) and developing a controller that uses *both* feedback and feedforward information to conduct transitions may lead to a more successful neuromorphic device. The described experiment was capable of operating only in closed loop mode that had the potential of being locked in a certain state if the sensory signals did not cross the necessary threshold. By applying underlying open loop control as well, it allows for a timing signal to prevent such an occurrence. This type of advanced SiCPG would be capable of operating with lower current stimulation methods such as intraspinal microstimulation (ISMS), to produce a low-power,

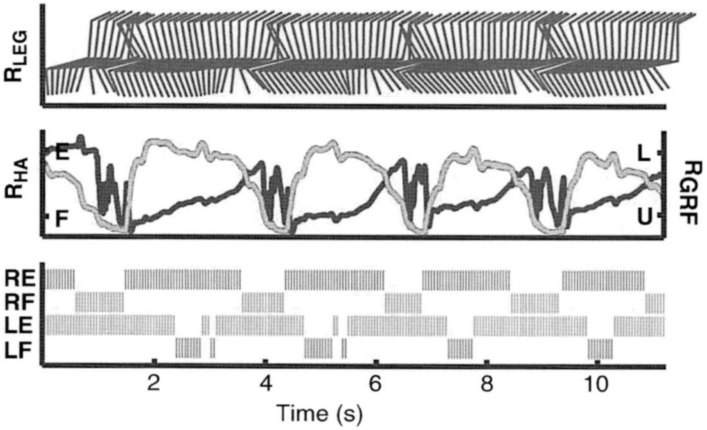

Figure 9.3 Results of one successful walking trial. The top trace is a stick figure representation of the hind leg movement over time. The second trace shows the right hip angle (HA) and ground reaction force. For the HA, it is possible to observe the changes from extension (E) to flexion (F). The GRF shows when the leg is loaded (L) or unloaded (U). The bottom trace shows the activation pattern from the SiCPG to the right and left extensors (E) and flexors (F). (Figure from Ref. [10].)

mobile neuroprosthesis. With further improvements and more detailed modeling of the biological CPG, an advanced and efficient neuromorphic hardware device for biohybrid control could be developed. The described application illustrates some of the capabilities of neuromorphic hardware for processing information from its environment to provide control of another system that in this case happens to be the activation of walking movements in an anesthetized cat.

9.2
Neuromorphic Hardware for Audition

The audition pathway is another complex biological system that has been emulated in integrated circuits. Specifically, there are implementations of models of the biological cochlea to process complex auditory signals in hardware [28–31]. While some of this material has been discussed in Chapter 3, a brief review of the basic components will be presented.

The biological cochlea acts as a filter bank for processing different auditory information [32]. The basilar membrane within the cochlea performs temporal and spatial computations that encodes information about the incoming auditory signal. Different locations along the membrane activate when different frequencies are present, allowing certain frequency information to pass while strongly rejecting others, and more specifically the higher frequencies relative to the current frequency tap. This spectral analysis translates the sound into spikes through the inner and outer hair cells, allowing the spiking information to be processed in higher-level neural circuits.

The functionality of the cochlea is implemented in silicon through the use of several band pass or low-pass filters with exponentially scaled cutoff frequencies similar to those observed in biology [33]. The reproducibility of the silicon cochlea improves as filter mismatch is corrected, instabilities are reduced, dynamic range improves, and inner and outer hair cell (IHC and OHC, respectively) models are incorporated. The frequency information from the biological cochlea is spaced out in an exponential manner as achieved in certain models.

Auditory applications involving neuromorphic hardware use the silicon cochlea as a front end to systems involving speech recognition, sound localization, and cochlear implants [6–8, 34–39]. Using a cochlear circuit allows for performing spectral computations in an analog domain (it outputs digital spikes with analog timing) that reduces the power consumption in comparison to a digital processor. This is especially important when designing cochlear implants because they operate on battery power and, as mentioned in Section 9.1, this allows for longer durations between changing batteries. It also allows for spectral information of the input signal to be computed in the analog domain and in real time as opposed to using a digital processor to perform computations on windows of information. Digital systems would require several frames of data to compute the spectral information that an analog system accomplishes in real time. In addition, each frame would require an analog-to-digital conversion resulting in more power consumed than when the computations were performed solely in the analog domain. It is possible to achieve a comparable performance rates in both analog and digital processors; however, the rate at which the conversions occur in the digital processor will result in a much higher power consumption rate.

The AER EAR is a neuromorphic device that uses two silicon cochlear circuits to provide spiking information through address event representation (AER, as discussed in Chapter 3) to another system for performing digital computations [7]. Typically used in multichip configurations, the AER EAR is useful when communicating with other devices with an AER interface. One control application involving the AER EAR uses the silicon cochlea's spectral computations to detect rhythmic sounds that in turn control the walking of a bipedal robot [39]. The following sections describe the specifics of the AER EAR and its use in such a control application.

9.2.1 AER EAR Architecture

The general building blocks of the AER EAR are two matched silicon cochlea circuits, inner hair cell (IHC) circuit models, and integrate-and-fire neurons with the AER interface [7]. Each silicon cochlea consists of 32 second-order low-pass filters using transconductance amplifiers operating in the weak regime (Figure 9.4). This allows the filters to operate while not requiring large amounts of current. The cutoff frequencies exponentially decrease and each filter produces an output voltage from the differential input values. The frequency range of the low-pass filters is from 50 Hz to 1 kHz.

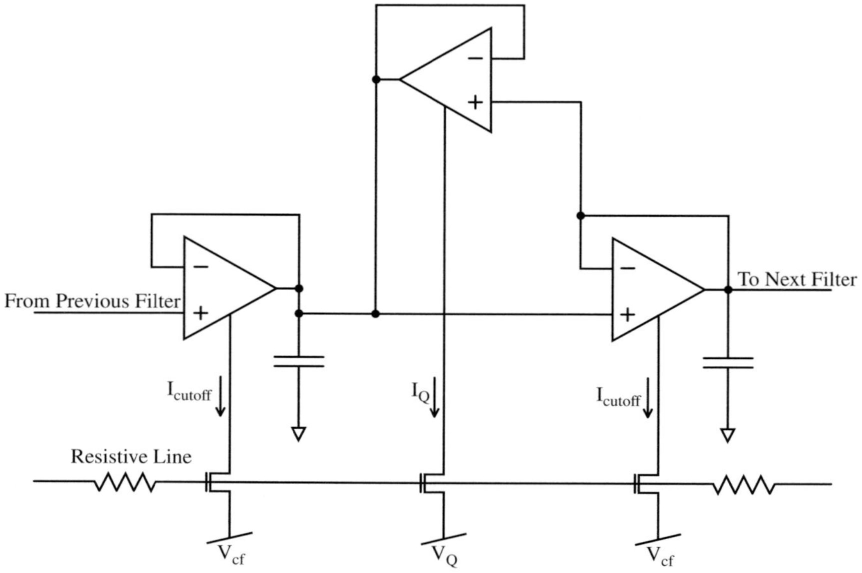

Figure 9.4 Schematic of one second-order low-pass filter. The resistive line allows for controlling the exponential spacing between cutoff frequencies. Each filter is connected to its neighboring filter and the voltage V_{cf} sets the cutoff frequency of this filter. The ratio of I_{cutoff} to I_Q also sets the Q factor of the filter. (Adapted from Ref. [7].)

The inner hair cell circuits are meant to emulate the biological function of transducing vibrations into neural signals. The differential voltage output from the cochlea circuit is passed into the IHC circuit and produces a single-ended output current. This output current is rectified and low-pass filtered before outputting to the neuron circuits.

The output current of the IHC is sent through a current mirror and passed to the membrane capacitance of the integrate-and-fire neuron. This current is integrated into the capacitor until the voltage crosses a certain threshold at which time the neuron produces a digital output to activate the AER circuitry. The AER output can be used by other AER communicating devices [7].

9.2.2
AER EAR Control Application

A control application involving the AER EAR uses the AER sound-based output from the silicon cochleas as the input to an AER-controlled audio filter [39]. This filter extracts the rhythm of the audio signal that controls the walking rate of a bipedal robot. The AER EAR front-end passes the spiking output to an AER filter. The output of the cochlea circuits contains information about the frequency of the input signal. This information is processed by the audio filter to extract the rhythm or "beat."

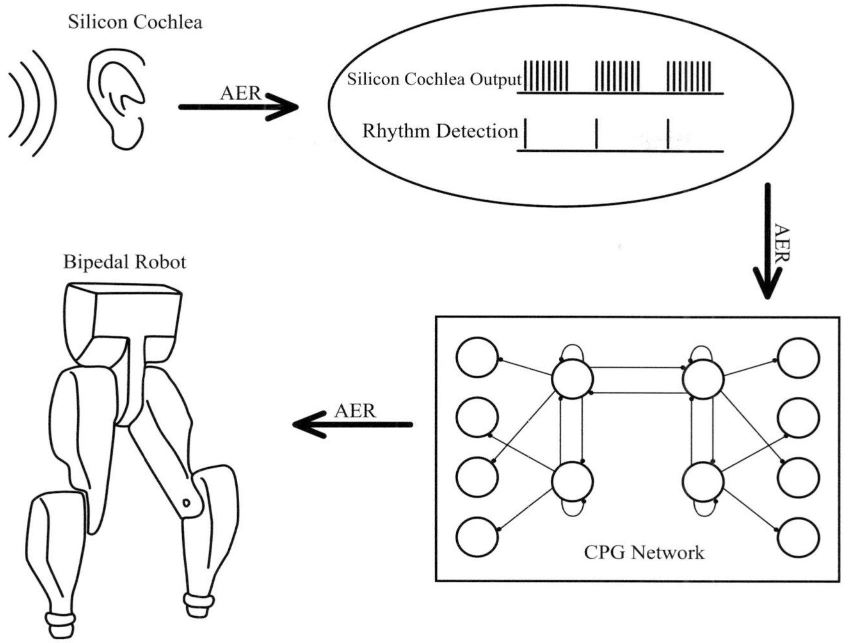

Figure 9.5 An audio signal is transmitted via AER into spikes to a rhythm detection circuit. This circuit produces activating pulses based on the rhythm of the cochlear input. The activating pulses coordinate the movement through the CPG40, which in turn controls the bipedal robot. (Adapted from Ref. [39].)

Beat detection is computed by determining the interspike interval (ISI) of certain cochlear output channels. The detection takes into account the spacing of the spikes to determine whether they are from the same burst of spikes or from repeating bursts. A programmed threshold determines the amount of time a burst of spikes will take in order to distinguish rhythmic beats. In this example, the beat threshold was set to 250 ms, which happens to be the amount of time required to make the bipedal robot walk on step. The rhythmic outputs from the AER filter are passed to an AER board with an FPGA onboard (Figure 9.5). The FPGA realizes a CPG network to operate the bipedal robot. The typical CPG network operating the robot consists of integrate-and-fire with adaptation neuron models and control hip flexion, hip extension, knee flexion, and knee extension in both left and right legs [40]. The output signals controlling the robot are driven by preprogrammed activation patterns. Each rhythmic output from the beat detector activates certain CPG commands programmed in VHDL in the FPGA. These commands activate certain actuators in the bipedal robot to generate walking at a certain rate. Synaptic weights of neuron models are no longer needed to adjust the walking cycle of the robot. The stepping frequency depends on the rhythm of the sound that allows for adjusting the walking speed by increasing the audio frequency. The outputs from the FPGA board are also transmitted via AER that allows for generating the desired pulses to the robots' actuators at the hips and knees of each leg. This output of pulses controls the specific activation pattern of hip and knee extensors and flexors generating the walking movements. This application demonstrates the ability of neuromorphic hardware to emulate multiple biological systems (in this case, audition and locomotion) to provide efficient control to an external system (the robot).

9.3 Neuromorphic Hardware for Vision

For neuromorphic vision applications, the front-end component is typically a model of the biological retina. This circuit is commonly called a silicon retina, which was the first sensory neuromorphic VLSI system by Mahowald and Mead [11], and attempts to replicate the functionality of the biological retina in terms of localized information processing. In order to emulate the biological retina, the circuit must be capable of encoding visual information in its output while operating over a wide dynamic range of input signals. Silicon retina circuits perform well as front-end transducers from light to electrical signal and have been described in detail in Chapter 3. Successful retinal circuits incorporate models of photoreceptors, horizontal cells, and bipolar cells or at least replicate their functionality [11, 41, 42]. The purpose of photoreceptor cells in biology is to transduce light into a voltage (represented by its membrane potential) [43]. The membrane potentials of several photoreceptor cells in the human retina are connected to horizontal cells that act as a low-pass spatial filter. The bipolar cells integrate the spatially low-passed signals from the horizontal cells and the photoreceptors to produce center-surround receptive fields. This information is transmitted to the visual cortex via the outer plexiform layer, ganglion cells, and the thalamus [43]. In some organisms, more complex image processing occurs at the retina. For example, rabbits can perform motion detection in the retina [44] and frogs perform orientation and scale discrimination [45]. These are functions necessary for their survival.

The localized signal processing occurring in the retinal cells is what makes it such a desirable biological process to emulate for other applications. Visual features incorporating spatial, orientation-based, and temporal filtering are computed at the pixel level and passed to the visual cortex. By incorporating this information, it enables performing processes such as edge detection and grouping or object tracking without having to process the scene in a frame-by-frame basis [2, 46, 47].

A simple example using a silicon retina imager demonstrates how edge tracking can control a motor [48, 49]. As a one-dimensional edge is shown to the sensor that has localized edge enhancement circuits at each pixel, the outputs of the pixel are used to compute the location of the edge. This voltage adjusts the position of a DC motor such that it rotates in the direction of the edge to keep it centered along the pixel array. This application in one dimension describes the efficiency a retina sensor could have on more complex control application (such as in two dimensions).

Other applications involving neuromorphic vision hardware include object detection [5] and object tracking [4, 49, 50]. In object detection, the retinal output is passed to an array of integrate-and-fire neurons where certain cells respond to different properties of the input image [5]. There are two types of cells in this application: simple and complex. The simple cell neurons respond to basic information about the image such as edge orientations. These simple cells transmit spiking information to complex cells that respond to more detailed shapes (such as 45 degree corners). Through the use of simple and complex cells, the network replicates the same computations believed to exist in the primary visual cortex (V1).

For object tracking, different approaches are taken to create a neuromorphic system, depending on what processes are emulated [4, 49]. In Ref. [49], the application represents a more detailed emulation of the biological counterpart that then requires programming a larger variable space. The system includes retinomorphic imaging, motion detection, object tracking, and selective attention modulation. The hardware incorporates spatial and temporal filtering of a visual scene in order to operate a motor that follows the object. Another approach uses a two-dimensional foveated retina designed for object tracking. In this case, the hardware emulates the biological functionality as opposed to the detailed processes emulated in Ref. [49]. This application will be discussed in further detail as the hardware can be used in applications involving smart surveillance or road following for an autonomous navigation system [3].

9.3.1 The Neuromorphic Imager (Silicon Retina)

The neuromorphic imager for tracking applications designed in Ref. [4] divides the pixel array into two sections: the fovea and the periphery. The pixel density ratio between these two regions is four, allowing for more detailed imaging in the foveal region. Another advantage of separating the image array into these two sections is that the peripheral pixels require less power than the foveal pixels, resulting in approximately four times less the power consumption than an array composed of foveal pixels alone. The computations in the foveal pixels emulate the biological function of smooth pursuit. In the periphery, the pixels emulate the saccadic movements in the primate ocular system. These two pixel types, when coupled with a motor system, provide the imager with the ability to either follow an object detected in the fovea region of the pixel array or perform saccades toward an object detected in the peripheral region.

The pixels in the fovea array perform localized computations to detect the velocity of an object in the field of view [4]. The velocity measurements are realized out of the array to reduce pixel sizes (and increase the fill factor). The foveal circuitry is depicted in Figure 9.6a where edge detection circuits are packed together in each pixel. The pixel operates on a log scale when exposed to low light and a linear scale in bright light. This is due to the current mirror configuration and sizing where the mirroring transistor current to voltage relationship is exponential during subthreshold operation and linear above threshold. The transistor attached to the

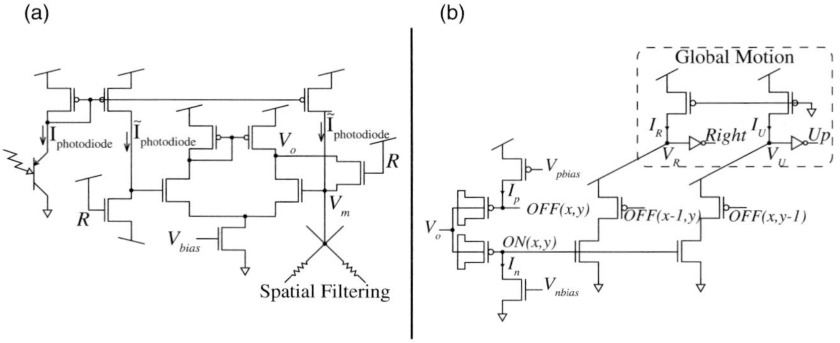

Figure 9.6 (a) Circuit schematic of the pixel that performs phototransduction, edge detection, and spatial filtering. (b) Motion detection circuit that determines the direction of motion the target is moving. (Adapted from Ref. [4].)

photodiode is also configured as a diode, thus operating in saturation mode, which allows for efficient gain control over wide dynamic ranges of the input. This prevents pixel saturation as the circuit scales the input appropriately to remain in the desired operating range.

The computations in the peripheral pixels search for a new object to track. Each peripheral circuit computes the centroid of the target object such that when the fovea decides to track it, the position is known such that it can be appropriately centered within the fovea array. High resolution is unnecessary for detecting a new target. This allows the use of larger photodiode sizes and fewer cells in the peripheral array. The edge detection and photoreceptor circuitry of these cells is similar to those of the fovea as in Figure 9.6a. One additional computation performed in the periphery is the detection of the arrival of new objects using a temporal ON-set detection circuit, depicted in Figure 9.6b. When a new object is detected, the motor system reorientates the retina so that the object can be examined by the high-resolution fovea, similar to what is observed in biology. In order to determine the location of the object, its centroid is computed using a resistance grid along the x- and y-directions. For robustness, spatial and temporal averaging of the centroid values is computed. This analog voltage models the mapping function of the superior colliculus, which is known to provide target locations for saccades in primates [51].

9.3.2 Visual Tracking

For tracking an object moving in the field of view, saccadic generation and smooth pursuit are emulated similar to what is observed in the primate ocular system [52]. The decision to saccade or track using smooth pursuit depends on the status of the fovea and peripheral circuits (Figure 9.7a). The peripheral pixels perform saccades by signifying a new target is available to track. The target is identified by peripheral pixels detecting a significant edge of an object whose intensity changes in time. The fovea region moves the target into the field of view. Once the object is located by the fovea

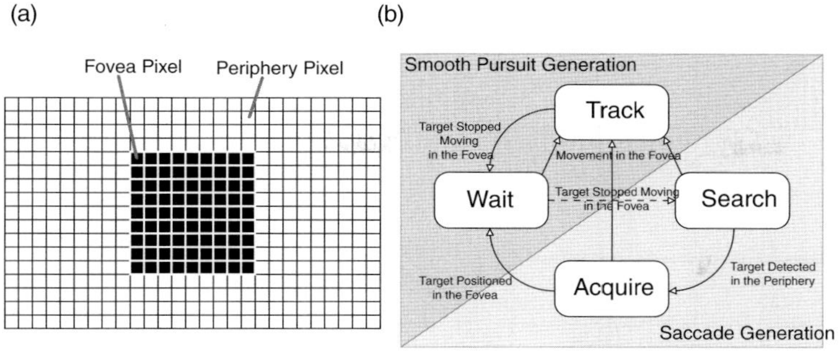

Figure 9.7 (a) The pixel layout of the neuromorphic imager. The fovea pixel cells are encompassed by the peripheral pixel cells. (b) The state machine controlling whether to saccade toward a new target found in the periphery or track an existing one detected in the fovea. (Adapted from Ref. [4].)

pixels, it attempts to keep it in the field of view by following the object's motion. This represents the smooth pursuit observed in biological visual systems. As previously mentioned, the motion detection is computed outside of the imaging array in order to save space and allow for a higher foveal pixel density.

The interactions between the peripheral and foveal pixels are determined by a state machine. There are four potential states that decide whether to generate saccades or smooth pursuit. This state machine is depicted in Figure 9.7b and runs as follows. If there is no motion in the fovea and an object is detected in the periphery, the position of the object is computed and the fovea is centered about that object. If the target moves within the foveal array, it is followed through smooth pursuit movements. If the followed object becomes stationary for a preset amount of time, smooth pursuit ceases and a new target search is generated in the periphery. If at any point motion is detected in the fovea while a target search is conducted in the periphery, the target search is abandoned and smooth pursuit starts in the fovea. These state decisions put emphasis on smooth pursuit while only performing saccade generations for a new target when nothing is detected or can be tracked in the fovea. If the preference is set to the saccadic movements as opposed to smooth pursuit, the imager would bounce around between different objects detected in the periphery and would never be able to completely track targets in the fovea.

9.3.3 Object Tracking Application

The completed hardware has pixel array sizes of 9×9 for the fovea and 19×17 for the periphery. The sensor detects and tracks people in a cluttered environment in both an indoor and an outdoor setting [4] (see Figure 9.8). Individuals running or walking from varying distances are tracked by the fovea array. One limitation to the system relates to the size of the fovea array. If the moving object is too far away, the low resolution in the fovea hinders its tracking ability because the motion is too small for the array to discern its location. Another setback to the fovea size is that fast moving

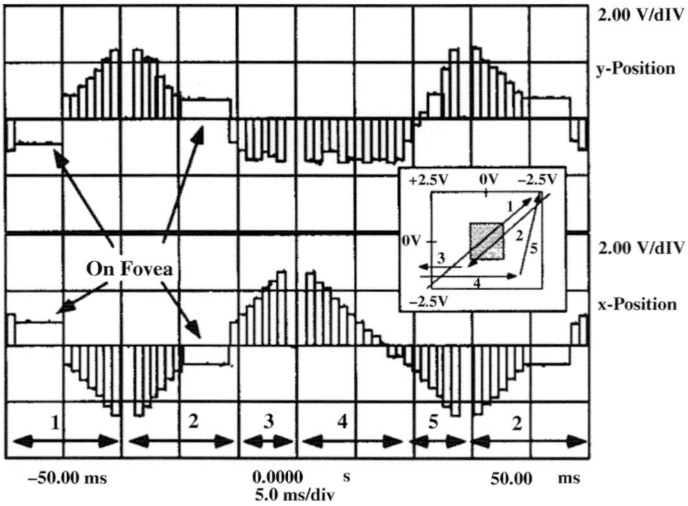

Figure 9.8 Oscilloscope trace of *x*- and *y*-positions of a high-contrast target moving across the pixel array as reported by the peripheral pixels. The *x*-direction is constant at 27 foveal pixels per second while the *y*-direction varies as specified by the numerical directions in the inset. When the target is over the fovea, the peripheral positions freeze at the last known state as shown on the graph (*On Fovea*). The inset represents the movement of the target across the entire pixel array as specified by the different numerical values. (Figure from Ref. [4].)

objects can move quickly out of the fovea's view faster than it can be tracked. When this happens, the periphery needs to redetect the target to center the fovea again and continue smooth pursuit.

A more complex application involving the sensor provides control to a robot for autonomous navigation [3] (see Figure 9.9). This application uses the neuromorphic imager as the front end of the system to provide coordinates to control the movements of a robot. The simple task of line following is accomplished by passing information about the state of the line to a microcontroller. The microcontroller uses these results to determine the required acceleration, deceleration, and steering to maintain an appropriate position along the line. To make the application more complex, obstacle avoidance is also incorporated to ensure the robot follows a specific path without hitting objects along the way.

In this application, two sensors are used to provide binocular vision to the robot; however, because the retinas' fields of view do not overlap, stereo depth perception is not implemented. Both the sensors attempt to maintain the line in a blind spot between the retinas' fields of view and essentially compete with one another to ensure the line be kept between them. The microcontroller has the ability to avoid obstacles in its path once detected. If an object is detected, the controller stops tracking the line and performs an avoidance maneuver to move around the said object. While navigating around the object, the controller remembers the avoidance direction such that it can remember how to reorient the robot back toward the line.

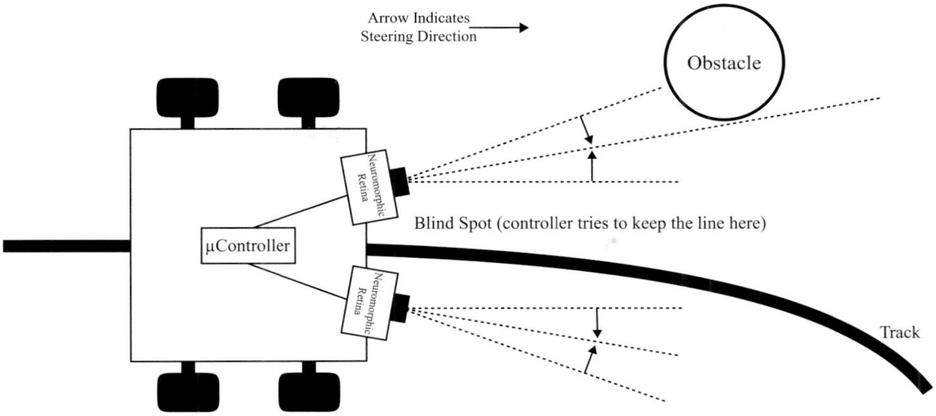

Figure 9.9 Block diagram depicting the autonomous navigation of the robot. The microcontroller receives information from the two neuromorphic sensors that determines the direction it should move to keep the line (or track) in the blind spot. In addition, it uses the information about incoming obstacles to determine when to steer away to avoid collisions. (Adapted from Ref. [3].)

The object avoidance always takes precedence over line tracking to ensure the robot does not collide with anything in its path.

As the robot navigates the path, the control strategy emphasizes accelerating when the path is straight and decelerating when the path curves or has a corner. The angle of the line relative to the *X*-axis position of the neuromorphic sensors determines the speed of the robot as well. If the angle of the line is less than 45 degrees, the *x*- and *y*-coordinates are small and large, respectively. These values correspond to decelerating the robot to ensure the line is maintained between the two sensors. The robot is successfully capable of traversing a path outlined by black tape on the gray linoleum floor. The lens on the sensor allows for a viewing angle of 10 degrees. The path contains black and white obstacles, two sharp turns, and some smooth “S” curves. As the robot moves at an average speed of 1 m/s it is capable of traversing the track with no errors. When the average speed increases, the errors observed occur from oversteering as opposed to problems with the tracking algorithm in the microcontroller.

9.4 Conclusions

Neuromorphic hardware possesses the ability to provide low-power control to systems such as robotics or neuroprosthetics. These devices are attractive because they perform computations in real time while using micro- to milliwatts of power. The applications described in this chapter are but a few of the ways neuromorphic hardware can be applied to control other external systems. As neuromorphic circuits become more compact and biologically realistic, more complex applications may arise utilizing their capabilities. In the future, biohybrid systems could incorporate neuromorphic hardware emulating vision, audition, proprioception, and locomotion in an effort to provide more complete control for robotics and living systems.

References

1 Mead, C. (1990) Neuromorphic electronic systems. *Proc. IEEE*, **78**, 1629–1636.

2 Choi, T., Shi, B., and Boahen, K. (2004) An on–off orientation selective address event representation image transceiver chip. *IEEE Trans. Circuits Syst. I*, **14**, 342–353.

3 Etienne-Cummings, R., Gruev, V., and Abdel-Ghani, M. (1999) VLSI implementation of motion centroid localization for autonomous navigation. *Adv. Neural Inf. Process. Syst.*, **11**, 685–691.

4 Etienne-Cummings, R., Van der Spiegel, J. Mueller, P., and Zhang, M. (2000) A foveated silicon retina for two-dimensional tracking. *IEEE Trans. Circuits Syst. II Analog. Digit. Signal Process.*, **47** (6), 504–517.

5 Folowosele, F., Vogelstein, J., and Etienne-Cummings, R. (2008) Real-time silicon implementation of V1 in hierarchical visual information processing, in Biomedical Circuits and Systems Conference, Baltimore, MD, pp. 181–184.

6 Baker, M.W. *et al.* (2003) A 16-channel analog VLSI processor for bionic ears and speech-recognition front ends, in Custom Integrated Circuits, San Jose, CA, pp. 521–526.

7 Chan, V., Liu, SC., and van Schaik, A.V. (2007) AER EAR: a matched silicon cochlea pair with address event representation interface. *IEEE Trans. Circuits Syst.*, **54** (1), 48–59.

8 Kumar, N., Himmelbauer, W. Cauwenberghs, G., and Andreou, A. (1998) An analog VLSI chip with asynchronous interface for auditory feature extraction. *IEEE Trans. Circuits Syst. II Analog Digit. Signal Process.*, **45** (5), 600–606.

9 Arena, P., Fortuna, L., and Frasca, M. (2002) Multi-template approach to realize central pattern generators for artificial locomotion control. *Int. J. Circ. Theor. App.*, **30** (4), 441–458.

10 Vogelstein, R.J., Tenore, F.V.G. Guevremont, L., Etienne-Cummings, R., and Mushahwar, V.K. (2008) A silicon central pattern generator controls locomotion *in vivo*. *IEEE Trans. Biomed. Circuits Syst.*, **2** (3), 212–222.

11 Mead, CA. and Mahowald, MA. (1988) A silicon model of early visual processing. *Neural Netw.*, **1** (1), 91–97.

12 Smith, L.S. and Hamilton, A. (eds) (1998) *Neuromorphic Systems: Engineering Silicon from Neurobiology, Progress in Neural Processing*, World Scientific Publishing Co. Ltd., London.

13 Mead F C. (1989) *Analog VLSI and Neural Systems*, Addison-Wesley Publishing Co., Reading, MA.

14 Dimitrijevic, M.R., Gerasimenko, Y., and Pinter, M.M. (1998) Evidence for a spinal central pattern generator in humans. *Ann. N. Y. Acad. Sci.*, **860**, 360–376.

15 Wilson, DM. (1961) The central nervous control of flight in a locust. *J. Exp. Biol.*, **38**, 471–490.

16 Bem, T., Orsal, D., and Cabelguen, J.M. (1993) Fictive locomotion in the adult thalamic rat. *Exp. Brain Res.*, **97** (2), 301–304.

17 Sillar, KT. and Roberts, A. (1998) A neuronal mechanism for sensory gating during locomotion in a vertebrate. *Nature*, **331**, 262–265.

18 Duysens, J. and Pearson, K.G. (1980) Inhibition of flexor burst generation by loading ankle extensor muscles in walking cats. *Brain Res.*, **187** (2), 321–332.

19 Cohen, AH. and Wallen, P. (1980) The neuronal correlate of locomotion in fish. *Exp. Brain Res.*, **41** (1), 11–18.

20 Grillner, S. and Zangger, P. (1979) On the central generation of locomotion in

the low spinal cat. *Exp. Brain Res.*, **34** (2), 241–261.

21 Russell, A., Orchard, G., and Etienne-Cummings, R. (2007) Configuring of spiking central pattern generator networks for bipedal walking using genetic algorithms, in International Symposium on Circuits and Systems, New Orleans, pp. 1525–1528.

22 Still, S., Hepp, K., and Douglas, R.J. (2006) Neuromorphic gait control. *IEEE Trans. Neural Netw.*, **17** (2), 496–508.

23 Tenore, F., Etienne-Cummings, R., and Lewis, M.A. (2004) A programmable array of silicon neurons for the control of legged locomotion, in International Symposium of Circuits and Systems, Vancouver, British Columbia, Canada, pp. 349–352.

24 Orchard, G., Russell, A., Mazurek, K., Tenore, F., and Etienne-Cummings, R. (2008) Configuring silicon neural networks using genetic algorithms, in International Symposium on Circuits and Systems, Seattle, pp. 1048–1051.

25 Chua, O. and Yang, L. (1998) Cellular neural networks: theory. *IEEE Trans. Circuits Syst. I*, **46** (2), 253–260.

26 National Spinal Cord Injury Statistical Center (February 2010) Facts and Figures at a Glance.

27 Baker, RJ. (2008) *CMOS: Circuit Design, Layout, and Simulation*, Revised 2nd edn (eds S.K. Tewksbury and J.E. Brewer), Wiley-Interscience, Hoboken.

28 Watts, F.L., Kerns, D.A. Lyon, R.F., and Mead, C.A. (1992) Improved implementation of the silicon cochlea. *IEEE J. Solid-State Circuits*, **27** (5), 692–700.

29 Lyon, RF. and Mead, CA. (1988) An analog electronic cochlea. *IEEE Trans. Med. Imaging*, **36** (7), 1119–1134.

30 van Schaik, A. and Meddis, R. (1999) Analog very large-scale integrated (VLSI) implementation of a model of amplitude-modulation sensitivity in the auditory brainstem. *J. Acoust. Soc. Am.*, **105** (2 Pt 1), 811–821.

31 Fragnière, E., van Schaik, A., and Vittoz, EA. (1997) Design of an analog VLSI model of an active cochlea. *Analog Integr. Circ. Signal Process.*, **13** (1–2), 19–35.

32 Rhode, W.S. and Greenberg, S. (1992) Physiology of the cochlear nuclei, in *The Mammalian Auditory Pathway: Neurophysiology* (eds A.N. Popper and R.R. Fay), Springer, New York, pp. 94–152.

33 Ruggero, M.A. (1992) Response to sound of the basilar membrane of the mammalian cochlea. *Curr. Opin. Neurobiol.*, **2**, 449–456.

34 Lande, T.S., Marienborg, J.T., and Berf, Y. (2000) Neuromorphic cochlea implants, in IEEE Proceedings on Circuits and Systems, Geneva, Switzerland, pp. 401–404.

35 Hamilton, T.J., Jin, C., van Schaik, A., and Tapson, J. (2008) An active 2-D silicon cochlea. *Biomed. Circuits Syst.*, **2** (1), 30–43.

36 Yu, T., Schwartz, A., Harris, J., Slaney, M., and Liu, S.C. (2009) Periodicity detection and localization using spike timing from the AER EAR, in International Symposium of Circuits and Systems, Taipei, Taiwan, pp. 109–112.

37 van Schaik, A. and Shamma, S. (2003) A neuromorphic sound localizer for a smart MEMS system, in International Symposium on Circuits and Systems, Bangkok, Thailand, pp. 864–867.

38 Lazzaro, J. and Wawrzynek, J. (1997) Speech recognition experiments with silicon auditory models. *Analog Integr. Circ. Signal Process.*, **13**, 37–51.

39 Gomez-Rodriguez, F. *et al.* (2007) AER auditory filtering and CPG for robot control, in International Symposium on Circuits and Systems, New Orleans, LA, pp. 1201–1204.

40 Lewis, M.A., Tenore, F., and Etienne-Cummings, R. (2005) CPG design using inhibitory networks, in International Conference on Robotics and Automation, Barcelona, Spain, pp. 3682–3687.

41 Boahen, K. and Andreou, A. (1991) A contrast sensitive silicon retina with reciprocal synapses, in *Advances in Neural Information Processing Systems* (eds J.E. Moody and R.P. Lippmann), Morgan Kaufman, San Mateo, CA, pp. 746–772.

42 Delbruck, T. (1993) Silicon retina with correlation-based velocity-tuned pixels. *Neural Netw.*, **4** (3), 529–541.

43 Dowling, J. (1987) *The Retina: An Approachable Part of the Brain*, Harvard University Press, Cambridge, MA.

44 Barlow, H.B. and Levick, W.R. (1965) The mechanism of directionally selective units in the rabbit's retina. *J. Physiology*, **178**, 447–504.

45 Lettvin, J.Y., Maturana, H.R., McCulloch, W.S., and Pitts, W.H. (1959) What the frog's eye tells the frog's brain. *Proc. Inst. Radio Eng.*, **47**, 1940–1951.

46 Liu, S.C. *et al.* (2001) Orientation-selective aVLSI spiking neurons. *Neural Netw.*, **32**, 629–643.

47 Venier, P., Mortara, A., Arreguit, X., and Vittoz, E. (1997) An integrated cortical layer for orientation enhancement. *IEEE J. Solid-State Circuits*, **32**, 177–186.

48 Indiveri, G. (1999) Neuromorphic analog VLSI sensor for visual tracking: circuits and application examples. *IEEE Trans. Circuits Syst. II*, **51**, 1337–1347.

49 Horiuchi, T.K. and Koch, C. (1999) Analog VLSI-based modeling of the primate oculomotor system. *Neural Comput.*, **11** (1), 243–265.

50 Okuno, H. and Yagi, T. (2008) A visually guided collision warning system with a neuromorphic architecture. *Neural Netw.*, **21** (10), 1431–1438.

51 Sparks, D., Lee, C., and Rohrer, W. (1990) Population coding of the direction, amplitude and velocity of saccadic eye movements by neurons in the superior colliculus. *Proc. Cold Spring Harb. Symp. Quant. Biol.*, **55**, 805–811.

52 Robinson, D. (1975) Oculomotor control signals, in *Basic Mechanisms of Ocular Motility and their Clinical Implications* (eds G. Iennerstrand and P. Bach-y-Rita), Pergamon Press, Oxford, pp. 337–374.

10
Biohybrid Systems for Neurocardiology

Peter H. Veltink, Lilian Kornet, Simone C.M.A. Ordelman, Richard Cornelussen, and Rik Buschman

10.1
Introduction

The heart is able to generate an activation and contraction pattern in a self-sustained manner. In a healthy heart, the activation starts at the sinoatrial node and is subsequently transmitted over the atria and ventricles to generate a well-coordinated contraction. The autonomic nervous system modulates this autonomous activation and contraction process, influencing heart rate (chronotropy), contractility of the heart (inotropy), and velocity of signal transduction in the AV node (dromotropy). It consists of two antagonistic parts: the ortho- and parasympathetic systems. In a healthy heart, generally speaking, the orthosympathetic system increases and the parasympathetic system suppresses cardiac activity.

A malfunctioning heart is, in many cases, life threatening. The heart rhythm can be too slow (bradycardia) and/or not synchronized in the ventricles or too fast (tachycardia). Bradycardia (a slow heart rate or missed heart beats) can be the result of an affected sinus node or an affected conduction system. The affected conduction system could be due to a disorder of the AV node, the His bundle, or the bundle branches. An affected conduction system could also result in asynchronous contraction of the ventricles. Artificial cardiac pacemakers can generate the missing activation pulses in case of bradycardia, synchronize atrium and ventricle in case of AV block, or optimize synchronization of the contraction of both ventricles in heart failure (cardiac resynchronization therapy, CRT [1–3]).

Too high heart frequencies (tachycardia) can be symptomatic or even life threatening if they result in fibrillation of the ventricles. Ventricular fibrillation might be due to the presence of a high sympathetic level in combination with, for example, heterogeneous innervation induced by a cardiac infarct or another reason of nerve deterioration [4]. Antiarrhythmic drugs such as those targeting ion channels have inadequate effectiveness and a risk of serious complications. To terminate ventricular tachyarrhythmias, implantable cardioverter defibrillators (ICDs) include antitachycardia pacing (ATP) and high-energy shocks (HES). ATP

Biohybrid Systems: Nerves, Interfaces, and Machines, First Edition. Edited by Ranu Jung.

effectively terminates 85–90% of ventricular tachycardia episodes minimizing the disadvantages of HES [5]. Instead of treating VF with high-energy shocks that reset the fibrillating heart, it would be better to suppress fibrillation by preventing, for example, high sympathetic levels. No clinically used stimulation system is available to perform this function. Present treatment using medication to lower sympathetic activity such as beta blockers continuously limit the capacity of the heart, thus preventing optimal heart function in the safe area of operation.

In recent years, stimulation of the parasympathetic innervation of the heart has been shown to effectively influence cardiac parameters. The aim of this chapter on biohybrid systems for neurocardiology is to provide an overview of this exciting field of neurocardiac monitoring and modulation, wherby a closed loop biohybrid system is formed. First, the main characteristics of the autonomic control of the heart are described. Subsequently, approaches in stimulation, sensing, and artificial control of the parasympathetic innervation of the heart are presented. Finally, conclusions are drawn and an outlook of future developments given.

10.2 Autonomic Neural Control of the Heart

The autonomic neural control of the heart consists of two antagonistic parts, which both consist of reflexive feedback loops and neural integration on several hierarchical levels.

10.2.1 Antagonistic Neural Control of the Heart

The autonomic nervous system modulates heart activity, which is autonomously generated by the heart itself. It consists of two antagonistic parts: the (ortho) sympathetic and the parasympathetic system [6]. Increased sympathetic activity results in increased cardiac activity: increased heart rate, increased contractility of the heart muscle, and decreased resistance of the large arteries (vasodilatation). Increased parasympathetic activity results in decreased cardiac activity, mainly decreased heart rate, and decreased conduction from the atrium to ventricle. At rest, parasympathetic activation dominates and heart rate is below the value, which would be obtained if the heart would be deprived of modulating neural inputs, the so-called intrinsic heart rate.

Parasympathetic preganglionic neurons originate from the brain stem and sacral spinal cord and synapse with short postganglionic neurons in terminal ganglia that are located near target organs [6]. The parasympathetic ganglia innervating the heart lie mainly embedded in epicardial fat, the so-called fat pads, on various locations on the atria and ventricles [7]. They are innervated by the 10th cranial nerve, the vagus, "wandering," nerve. The sympathetic system has paravertebral ganglia close to the spinal cord and prevertebral intrathoracic ganglia. Preganglionic sympathetic neurons that influence heart function originate at thoracic levels from the spinal cord [6]

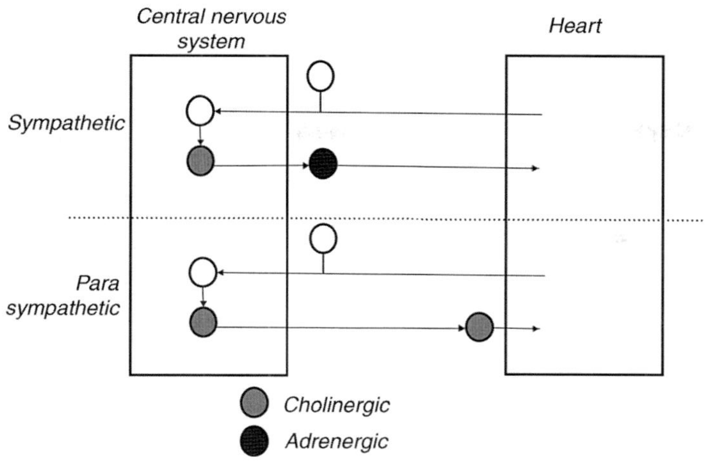

Figure 10.1 Sympathetic and parasympathetic systems reflexively control the heart. Ganglia of the sympathetic system lie close to the spinal cord and those of the parasympathetic system lie epicardially. Neurotransmitters released by the pre- and postganglionic neurons are partly different, as indicated.

and mostly synapse with long postganglionic neurons in these ganglia. However, preganglionic sympathetic nerves were also found to synapse in ganglia on the heart [8]. Preganglionic and parasympathetic postganglionic axons involved in neural control of the heart release the neurotransmitter acetylcholine and are therefore called cholinergic. The sympathetic postganglionic axons release norepinephrine and are called adrenergic (Figure 10.1).

10.2.2 Reflexive Neural Control of the Heart

Both ortho- and parasympathetic systems are reflexive in nature (Figure 10.1). These reflexes provide feedback control of the neural modulation of the heart. Sensors providing feedback are mainly mechano- and chemosensitive. Stress-sensitive baroreceptors are present in both the high- and the low-pressure parts of the circulation. In the high-pressure system, the main baroreceptors are located in the carotid sinus and aortic arch. Baroreceptors exhibit static and to some extent rate sensitivity, and generate signals that are predominantly phasic in nature and in synchrony with the heart cycle [9–11]. They provide feedback for the primary regulation of the aortic pressure through the *baroreceptor reflex*. The afferent information of the carotic sinus receptors is carried by the sinus nerve, which join the glossopharyngeal nerve [6]. The afferents from the aortic arch receptors are included in the depressor branch of the vagus nerve [6]. Low-pressure sensors are located in the junction of the atria with their veins, the pulmonary artery and the endo- and epicardium of the atria and ventricles [6]. Stress measured by these sensors depends mainly on venous return. Therefore, these sensors are important in sensing blood volume and are involved in the *Bainbridge reflex*, which stimulates heart rate and contractility if blood volume at the venous side of the circulation increases.

In addition to controlling blood pressure, the neural regulation of the heart depends on oxygen and carbon dioxide concentrations, which are sensed by chemoreceptors. Such sensors are located both peripherally near the baroreceptors in the carotid sinus and aortic arch and centrally in the medulla [6]. The peripheral chemoreceptors predominantly sense arterial oxygen pressure, while the central receptors sense pH, related to arterial carbon dioxide pressure. In addition, mixed mechano- and chemosensitive sensors are present in the walls of all cardiac chambers [12]. Their information is transmitted by unmyelinated fibers in the vagus nerve. These mixed sensors are involved in the *Bezold–Jarisch reflex*, activation of which causes inhibition of efferent sympathetic activity, reducing atrial blood pressure [12]. The baroreceptor and Bezold–Jarisch reflexes are complementary in blood pressure control. During normal conditions, the baroreceptor reflex is dominant.

10.2.3
Hierarchical Neural Control of the Heart

In addition to the feedback control of heart contraction via reflexive loops to the brain stem, an extensive neural network is present peripheral to the brain, consisting of interconnected ganglionated plexi on the epicardium of the heart. This complex neural network on the heart, also called "the little brain on the heart" integrates afferent and efferent as well as parasympathetic and sympathetic activity as a final relay station, and is thought to be important in the coordinated neural control of the heart [8, 13]. In addition to direct neural modulation, the cardiac system is also more systemically modulated by catecholamines circulating in the blood, for example, released by the adrenal medulla.

In conclusion, the cardiovascular system is modulated by a complex distributed multilevel reflexive neural system, consisting of two antagonistic parts.

10.3
Monitoring and Modulating the Autonomic Reflexive Control of the Heart

In addition to direct cardiac pacing, artificial stimulation of the autonomic neural system may provide many opportunities for providing support in case of cardiovascular disorders. Related to its function, stimulation of the sympathetic system is expected to result in increased cardiac output, involving increased heart rate and stronger contraction of the heart and vasoconstriction, resulting in a higher blood pressure. In contrast, stimulation of the parasympathetic system is expected to result in decreased cardiac output. The latter is especially relevant since conventional cardiac pacing does not allow reducing cardiac activity, which is important if the load on a weak heart becomes too high. By reducing load, parasympathetic stimulation is expected to be able to reduce the occurrence of tachycardia in heart failure. Also, increased parasympathetic activation may limit the sympathetic activity, which in remodeled or sick hearts, like ischemic hearts or those with a previous cardiac infarct [4], might be heterogeneous, imposing a risk for the induction of arrhythmias. Both, that is, reducing load and inhibiting sympathetic activation, may be important in avoiding ventricular fibrillation, which is life threatening. In addition, parasympathetic stimulation might influence blood flow within the cardiac tissue, lungs, and other organs and have a positive effect on inflammation markers present in heart failure patients [14].

In order to control stimulation of the autonomic system adequately, relevant parameters describing the state of the cardiovascular system need to be sensed. For example, the following parameters may be determined and collected using the electrical activity of the patient's heart: heart rate, heart rate variability, heart rate turbulence, T-wave alterations, P-P, P-R, A-V intervals, S-T elevation, and so on. In addition, artificial sensors for measuring blood pressure-derived parameters or lung fluid status and chemical compounds can be used such as partial oxygen pressure, brain natriuretic peptide content, and pH. If these indicate that the status of the heart has become dangerously unbalanced, stimulation of the parasympathetic system can be initiated. Alternatively, information can be derived about the chemical or pressure state by sensing information conducted by the autonomic nerves. The question is whether we can derive these signals selectively from these nerves, which carry afferent and efferent signals for many body functions.

When taking a closer look, the problem of artificial neural modulation of the heart is complex and challenging, considering that we place an artificial closed loop control system, including sensing and stimulation, in parallel with a dysfunctioning cardiovascular system under multilevel closed loop neural control (Figure 10.2). The challenge then in is to develop an appropriate design of an artificial support system in order for the combined biohybrid system to function optimally.

The following sections will provide an overview of the state of the art on sensing, stimulation, and artificial neural control of the cardiovascular system. First, the

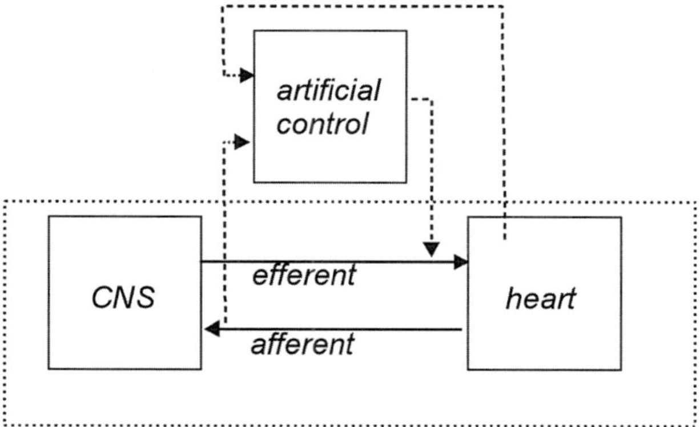

Figure 10.2 Schematic representation of an artificial control system to monitor and modulate cardiac function, placed in parallel to the dysfunctioning neurocardiac system.

possibility of deriving information from the mechanical and chemical sensors of the cardiovascular system will be discussed. Subsequently, two approaches taken in the artificial stimulation of the parasympathetic neural control system of the heart will be presented: stimulation of the vagus nerve and stimulation of the parasympathetic ganglia or nerves on the heart. Finally, the first studies on a biohybrid system consisting of an artificial closed loop neural control of the heart will be presented.

10.3.1
Sensing of Afferent Signals from and Efferent Signals to the Heart

Zucker and Gilmore recorded afferent discharges of individual receptors in the atria from single nerve fibers of the cervical vagus in dogs. During normal heart function, these receptors exhibited a rhythmic discharge pattern in synchrony with the heart cycle [11].

The recordings performed by Zucker and Gilmore require dissection of the vagus nerve, which can be performed only during acute experiments. Less-invasive recording techniques are required if signals are to be recorded chronically. Several electrode configurations have been proposed, which potentially can be used for chronic recording. Whole-nerve recordings can be obtained from tripolar electrode configurations inside an insulating cuff that are placed around the nerve [15–18]. Recently, we have evaluated whole-nerve recording from the vagus nerve of pigs. In 10 out of 12 pigs, we found a modulation of the signal with the heart cycle after ensemble averaging the nerve signals over approximately 100 cycles using an electrocardiogram (ECG) to identify the cycles (Figure 10.3). The need for ensemble averaging over many heart cycles lies in the fact that the cyclically active sensory fibers from the cardiac system constitute only a relatively small percentage of all fibers in the vagus nerve trunk. They are, therefore, hard to be selectively sensed from the outside. The major problem of whole-nerve recordings is therefore selectivity. As described, the vagus nerve carries bidirectional information for control and monitoring of many body functions, including cardiac and visceral. Efferent and afferent myelinated and unmyelinated nerve fibers involved in these different functions lie partly organized in bundles, the so-called fascicles, inside the vagus nerve. Fibers that transmit information about pressure or chemical compounds from the heart to the central nervous system have a limited range of fiber diameters and therefore limited range of transduction velocities. To selectively record certain signals in these afferent fibers, we will need a velocity-selective method. Donaldson *et al.* presented a method for velocity-selective, and thus fiber diameter-selective, whole-nerve recordings using a cuff configuration with many electrodes positioned along the nerve [19, 20], but did not apply their method to vagus nerve recording. In this method, delayed signals on subsequent electrode tripoles are added, thus amplifying signals that propagate with a certain velocity. This method can also distinguish between afferent and efferent signals, but requires a long cuff with many electrodes along the nerve. Improved spatial selectivity can be obtained using intrafascicular electrodes. For this purpose, longitudinal intrafascicular electrodes (LIFEs) were developed [21–24], but they have not yet been applied for vagus nerve recording. The low-conducting connective tissue

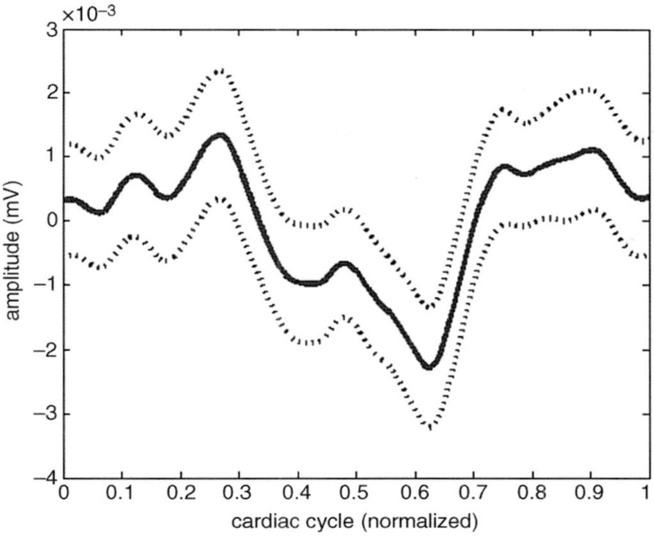

Figure 10.3 Example registration from a pig experiment showing modulation of the whole nerve vagal signal with the heart cycle after ensemble averaging the nerve signals over approximately 100 cycles. The signals of all cycles were centered on the R-tops of the ECG. The solid line represents the mean and the dotted lines the mean ±95% confidence interval.

sheet around each fascicle, called perineurium, provides insulation from signals from other fascicles, thus having the same function as an insulating cuff for extraneural recording. Intrafascicular electrodes have been reported to be biocompatible and cause little nerve damage [23]. The problem with intrafascicular recording in a mixed nerve with many fascicles is to identify the fascicles involved in the targeted function, in our case neurocardiac control and its invasiveness with potential detrimental effects in a chronic setting.

Evoked responses, called nerve compound action potentials (NCAPs) can be derived after electrical stimuli of the vagal nerve using extraneural tripolar nerve cuff electrode configurations [25–27]. These NCAPs consist of several peaks with varying latencies, each being generated by a different class of excited nerve fibers: the peak with the smallest latency is generated by the thickest myelinated Aβ fibers having highest conduction velocity. Subsequent peaks represent the slower and thinner myelinated nerve fibers (Aγ and B fibers) and finally unmyelinated fibers. In general, the fibers with higher conduction velocities have a lower threshold. Therefore, at progressively higher stimulation levels, additional peaks with larger latencies appear in the NCAP.

Ordelman *et al.* [25] showed that an evoked fast-traveling, relatively dispersed long-latency component coming from the periphery is present in the NCAP when applying the average reference recording method (Figure 10.4). This method records the signal of each electrode relative to the average signal of all electrodes. The fast-traveling long-latency component is not present in the usual tripolar nerve recording,

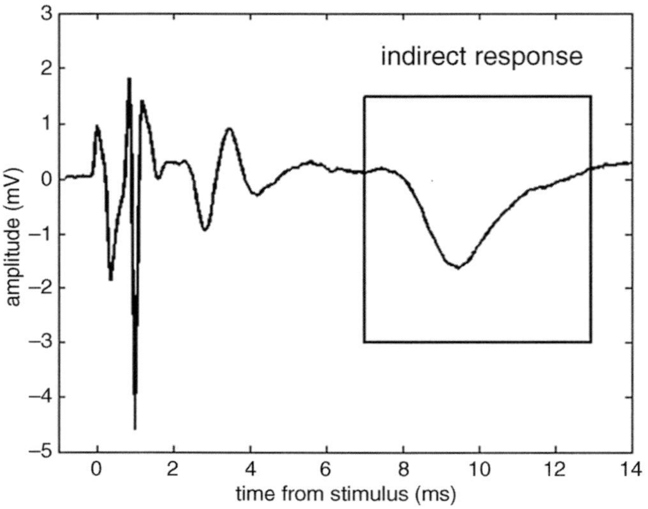

Figure 10.4 Example a compound action potential measured from the vagal nerve, showing an indirect long-latency component, which was shown to come from the periphery and to have a relatively high conduction velocity [25]. The experiments were performed on pigs.

due to the second order spatial filtering, which is implicit in this method. The latency of this long-latency component changed when influencing the state of the heart using high-frequency stimulation and is, therefore, hypothesized to be influenced, at least in part, by the state of the heart and potentially generated by the "little brain on the heart" [8, 25]. Further investigations have to validate this hypothesis and investigate whether this NCAP can provide useful information about the state of the heart to be used as a monitoring index and for feedback in the control of vagus nerve stimulation.

10.3.2 Stimulation of the Parasympathetic Input to the Heart

Neural stimulation of the parasympathetic system has been shown to have negative chronotropic and dromotopic effects [28–32] and to be potentially protective in heart failure [33]. Several approaches have been taken to stimulate the parasympathetic innervation of the heart. The first method stimulates the vagus nerve on the cervical level with electrodes on the nerve or inside the superior vena cava [30]. Alternatively, the ganglionated plexuses on the heart can be stimulated from within the heart [30, 31] or with epicardial electrodes on the heart [32]. Both approaches will be described separately in the following sections.

10.3.2.1 Stimulation of the Cervical Vagus Nerve

Stimulation of the vagus nerve at the cervical level is a common clinical practice in refractory epilepsy patients [34–37]. The cervical vagus nerve can be approached easily during surgery. Besides reduction of seizure frequency in epilepsy through activation of afferent fibers, stimulation of the vagus nerve can have great effects on

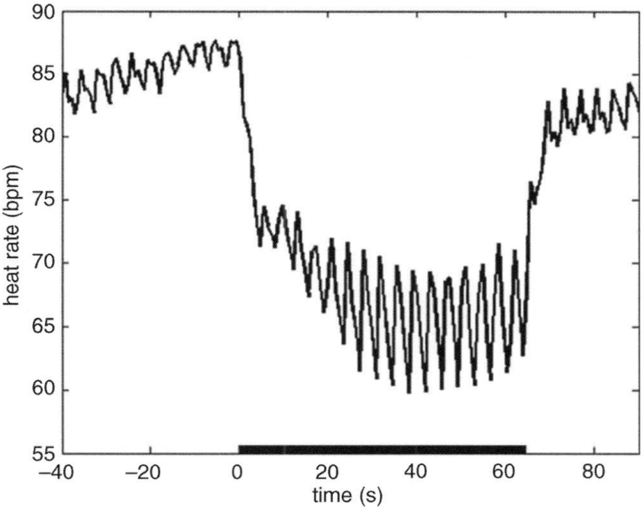

Figure 10.5 Stimulation of the vagal nerve can reduce the heart rate: example recording from a pig experiment. The bar at the bottom of the graph indicates the period during which the vagal nerve was stimulated. The alternation of the heart rate is due to breathing.

heart rhythm (Figure 10.5), can delay the atrial–ventricular conduction, and even lead to blocking of atrial–ventricular conduction (Figure 10.6). During vagus nerve stimulation for epilepsy, cardiac effects are avoided by stimulating the left vagus nerve, which is thought to have a relatively low effect on heart rate, and by applying

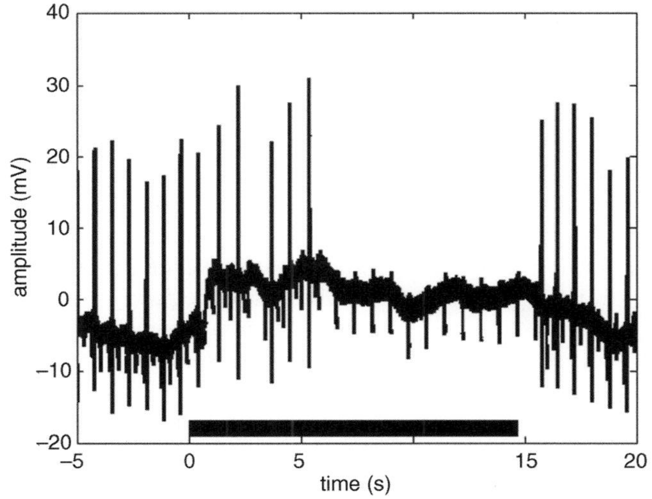

Figure 10.6 Stimulation of the vagal nerve can lead to blocking of atrial–ventricular conduction (AV block): example recording from a pig experiment. The black bar at the bottom of the figure indicates the period in which the vagal nerve was stimulated. The displayed ECG recording shows an absence of QRS complexes after a number of seconds, indicating an absence of ventricular activation due to an AV block.

relatively low stimulation amplitudes. Therefore, adverse cardiac effects occur rarely, when stimulating the vagus nerve for treatment of epilepsy [36, 38]. In recent years, vagus nerve stimulation, with the purpose of influencing the heart, has been investigated [27, 33, 39]. Schwartz *et al.* [33] studied feasibility, safety, and possible efficacy of chronic vagus nerve stimulation on the evolution of heart failure in eight heart failure patients with a follow-up of 6 months. The right vagus nerve was chronically stimulated using an asymmetrical bipolar multicontact cuff with cathodic excitation and subsequent unilateral anodal block, which was assumed to preferentially activate efferent fibers. Primarily, it was found that the treatment is feasible, safe, and tolerable, with only minor side effects such as hoarseness, coughing, and sensation of stimulation. In addition, this small study showed a significant improvement in quality of life, NYHA functional class, and left-ventricular end-systolic volume. Definite conclusions regarding efficacy need to follow from a larger placebo–controlled randomized study in humans. Zhang *et al.* [39] further investigated chronic vagus nerve stimulation in a randomized control study in 15 dogs with heart failure induced by high-rate ventricular pacing. The left ventricular ejection fraction, heart rate variability, and baroreflex sensitivity were significantly larger after chronic vagus nerve stimulation than if none was applied. In addition, pronounced antiinflammatory effects of chronic vagus nerve stimulation were found.

An important practical side effect of vagus nerve stimulation is hoarseness [36], which is associated with activation of the large diameter myelinated efferent fibers innervating the vocal folds through the recurrent laryngeal nerve. This nerve branches from the vagus nerve caudally from the site where the vagus nerve is surgically accessible for placing stimulation electrodes. These large-diameter myelinated fibers have a relatively low excitation threshold because of their size [40]. Tosato *et al.* recently proposed to selectively block the activation of intrinsic laryngeal muscles during vagus nerve stimulation using quasitrapezoidal pulses applied in a tripolar electrode cuff configuration [27]. They demonstrated 75% reduction in the first component of the NCAP, associated with major reduction in the excitation of the thickest myelinated nerve fibers and the intrinsic laryngeal muscles. This method of anodal blocking had been previously demonstrated by Fang and Mortimer [41], who did not apply the method to selective stimulation of the vagus nerve. It was compared with alternative methods of diameter-selective activation such as depolarizing prepulses and slowly rising stimulation pulses by Vuckovic *et al.* [42].

An alternative method for selective activation of cardiac fibers in the vagus nerve may be intrafascicular stimulation [43–48]; however, this method has not been applied to vagus nerve stimulation as far as we know. The problem with selective stimulation using intrafascicular electrodes is how to identify the fascicles involved in the targeted function, in our case neurocardiac control. More important, it is an invasive approach, which may not be suited for use in patients. Finally, stimulation of cardiac branches of the vagus nerve may be an option. However, it may be more difficult to approach them because they might appear relatively caudally and they are relatively small compared to the vagal trunk [49].

10.3.2.2 Stimulation of the Autonomic Cardiac Innervation

Chen *et al.* [28] and Quan *et al.* [29] demonstrated that intracardiac stimulation of the nerves innervating the AV node in the right postero- and anteroseptal areas could induce slowing of the ventricular rate and a complete AV conduction block in acute human experiments. In both acute studies, stimulation was performed during spontaneous or induced atrial fibrillation in order to avoid activation of the myocardium. Alternatively, Schauerte *et al.* [30] avoided stimulation of the myocardium by using high-frequency burst stimulation, applied during the atrial refractory period in animals. Bianchi *et al.* demonstrated that negative dromotropic effects could also be obtained using a standard atrial pacing lead at a septal position [31]. Besides using the intracardiac septal position [50], Schauerte *et al.* demonstrated negative dromotropic effects using intracardiac electrodes, placed in the inferior vena cava [51] or coronary sinus [30]. More recently, several studies reproduced and further investigated the induction of negative dromotropic effects by stimulation of fat pads. In an acute dog study, a lead was fixated epicardially into parasympathetic neural structures that supply the AV node to decrease the ventricular rate during AF [52]. In a chronic dog study, Zhang *et al.* demonstrated that regular slow ventricular rhythm could be obtained during atrial fibrillation by combining epicardial AV node stimulation with ventricular pacing [53], and that chronic AV node stimulation using a epicardial lead on the AV node fat pad resulted in beneficial long-term ventricular rate slowing [54]. Rossi *et al.* [32] used a temporary heart wire, implanted for 5 days in 32 patients during bypass surgery, showing complete AV block in 29 patients and reduced ventricular rate during atrial fibrillation in 14 patients. Chronic effects of AV node stimulation on ventricular rate during AF has been found by us for up to 8 months (unpublished) and others for upto 2 years [50].

Future alternatives to or improvements of AV node stimulation for treatment of ventricular arrhythmias in heart failure may be based on an improved understanding of possible causal relations between sympathetic and parasympathetic inputs to the heart and these arrhythmias [55]. Several recent studies provide relevant knowledge that has not yet been applied in new therapies. Chiou *et al.* [56] demonstrated in patients with AV node-related tachycardia that increased vagal tone can induce tachycardia because it has different effects on the refractoriness of slow and fast AV nodal conduction pathways. In addition, Ogawa *et al.* [57] and Zhou *et al.* [58] demonstrated relations between specific patterns of autonomic nerve discharge and cardiac arrhythmias during heart failure in which sympathetic input patterns play a major role. They showed that the majority of malignant ventricular arrhythmias are preceded by increased sympathetic input, having specific patterns, including low-amplitude burst and high-amplitude spike discharges [58]. Also, the reduction in sympathovagal balance at night appeared to be due to reduced sympathetic discharge rather than a net increase in vagal activity [57]. Swissa *et al.* demonstrated that long-term subthreshold stimulation of the left stellate ganglion induces sympathetic and parasympathetic sprouting and, consequently, hyperinnervation in both right and left atria in dogs [59]. Lewis *et al.* demonstrated that vagus nerve stimulation decreases left ventricular contractility involving a degree of interaction between vagal and sympathetic influences [60]. They paced the heart in order to avoid the confounding effect of heart rate changes. These new insights have yet to be used in improved therapies.

10.3.3
Biohybrid Closed Loop Artificial Neural Control of the Heart

Research on closed loop control of cardiac function involving stimulation of the autonomic neural system has been limited until now. Tosato *et al.* demonstrated closed loop control of heart rate by electrical stimulation of the vagus nerve in pigs [26], sensing heart rate from ECG. Heart rate could be changed maximally 20% with respect to baseline. The reason for this limited range of operation was suggested to be the parallel action of physiological reflexes. These reflexes may have been stimulated directly or reacted indirectly to the effect of the vagal nerve stimulation. In fact, artificial heart rate control was demonstrated to be able to overrule this parallel physiological control system in the above-mentioned limited range. A different approach to closed loop artificial control of the heart was taken by Sato *et al.* [61–63], who replaced a nonfunctioning baroreceptor reflex in baroreflex failure rats by an artificial bionic baroreflex. This artificial reflex was called bionic because it was implemented using a computerized system, simulating the dynamics of the vasomotor center in real time, and artificially closing the interrupted baroreceptor loop by providing the systemic blood pressure, as sensed by a micromanometer, as the input and by stimulating the sympathetic efferent nerves based on the output of the simulated vasomotor center dynamics. The open-loop dynamics of the physiological baroreflex system was first identified by system identification methods using white noise perturbations. After implementing the bionic baroreflex, blood pressure responses to head-up tilt were demonstrated to be comparable to intact baroreflex responses. Artificial replacement of a dysfunctioning baroreflex may be relevant in patients with neurological disorders that lead to baroreflex failure, such as traumatic spinal cord injuries and baroreceptor deafferentation [61]. It should be noted, however, that these principles have been demonstrated only in acute animal experiments and robust chronic implementation may be challenging for several reasons, including the need for a reliable chronically implanted sensor for blood pressure.

Biohybrid closed loop artificial neural control of the heart is still a big challenge since it needs to be integrated with the affected neural control of the heart, of which the complex multilevel dynamical characteristics still remain largely unknown and which may be only partially observable with the limited sensory information that can be practically measured. In general, the remaining physiological neural feedback control and the added artificial feedback control will interfere in a manner that may be understood only partly with available knowledge. For this reason, the first feasible approach beyond open-loop stimulation of the neurocardiac system, which is being tested in humans as reported in Section 10.3.2, is expected to be conditional neural stimulation if the status of the heart, as monitored with feasible means, is detected to be dangerously unbalanced and predicted to result in a hazardous state of the heart. In this manner, parasympathetic stimulation could be initiated if atrial or ventricular fibrillation is predicted based on information sensed from the cardiac and neural system.

10.4 Conclusions

Artificial neuromodulation of the cardiac system by stimulation of the autonomic neural system complements conventional cardiac pacing. The clinical potential of stimulation of cervical vagus nerve, its cardiac branches, or the neural plexuses on the heart using intracardiac leads has been demonstrated in recent years [30–33]. However, important questions remain regarding artificial neural sensing, stimulation, and control. The essential issue in neural sensing and stimulation is to achieve adequate selectivity using interface methods, which are potentially clinically feasible. The challenge is to design the artificial control system such that the combined physiological and artificial control is functioning in an optimal manner. This requires a better understanding of the dynamics of the complex autonomous neural control of the heart, improved observability of this complex system using adequate sensing, and good controllability. At present, several open-loop approaches for stimulation of the autonomous neural control of the heart are investigated in human studies. The next step is to apply these methods conditionally upon detection or better, prediction of hazardous states of the heart.

References

1 Cleland, J.G.F., Daubert, J.C., Erdmann, E., Freemantle, N., Gras, D., Kappenberger, L., Klein, W., and Tavazzi, L. (2001) The CARE-HF study (cardiac resynchronization in the heart failure study): rationale, design and end-points. *Eur. J. Heart Fail*, **3**, 481–489.

2 Cleland, J.G.F., Daubert, J.C., Erdmann, E., Freemantle, N., Gras, D., Kappenberger, L., and Tavazzi, L. (2005) The effect of cardiac resynchronization on morbidity and mortality in heart failure. *N. Engl. J. Med.*, **352**, 1539–1549.

3 Lim, S.H., Lip, G.Y.H., and Sanderson, J.E. (2008) Ventricular optimization of biventricular pacing: a systematic review. *Europace*, **10**, 901–906.

4 Zipes, D.P. (2008) Heart–brain interactions in cardiac arrhythmias: role of the autonomic nervous system. *Cleve. Clin. J. Med.*, **75** (Suppl 2), S94–S96.

5 Sweeney, M.O., Sherfesee, L., DeGroot, P.J., Wathen, M.S., and Wilkoff, B.L. (2010) Differences in effects of electrical therapy type for ventricular arrhythmias on mortality in implantable cardioverter-defibrillator patients. *Heart Rhythm*, **7**, 353–360.

6 Boron, W.F. and Boulpaep, E.L. (2003) *Medical Physiology*, 1st edn, Saunders.

7 Armour, J.A., Murphy, D.A., Yuan, B.X., MacDonald, S., and Hopkins, D.A. (1997) Gross and microscopic anatomy of the human intrinsic cardiac nervous system. *Anat. Rec.*, **247**, 289–298.

8 Armour, J.A. (2007) The little brain on the heart. *Clev. Clin. J. Med.*, **74**, S48–S51.

9 Armour, J.A. and Ardell, J.L. (eds) (2004) *Basic and Clinical Neurocardiology*, Oxford University Press.

10 Arndt, J.O., Dorrenhaus, A., and Wiecken, H. (1975) The aortic arch baroreceptor response to static and dynamic stretches in an isolated aorta-depressor nerve preparation of cats *in vitro*. *J. Physiol.*, **252**, 59–78.

11 Zucker, I.H. and Gilmore, J.P. (1973) Left atrial receptor discharge during atrial arrhythmias in the dog. *Circ. Res.*, **38**, 15–19.

12 Campagna, J.A. and Carter, C. (2005) Clinical relevance of the Bezold–Jarisch reflex. *Anesthesiology*, **98**, 1250–1260.

13 Armour, J.A. (2008) Potential clinical relevance of the "little brain" on the mammalian heart. *Exp. Physiol.*, **93**, 165–176.
14 Tracey, K.J. (2002) The inflammatory reflex. *Nature*, **420**, 853–859.
15 Andreasen, L.N. and Struijk, J.J. (2002) Signal strength versus cuff length in nerve cuff electrode recordings. *IEEE Trans. Biomed. Eng.*, **49**, 1045–1050.
16 Andreasen, L.N., Struijk, J.J., and Lawrence, S. (2000) Measurement of the performance of nerve cuff electrodes for recording. *Med. Biol. Eng. Comput.*, **38**, 447–453.
17 Struijk, J.J. (1997) The extracellular potential of a myelinated nerve fiber in an unbounded medium and in nerve cuff models. *Biophys. J.*, **72**, 2457–2469.
18 Struijk, J.J., Thomsen, M., Larsen, J.O., and Sinkjaer, T. (1999) Cuff electrodes for long-term recording of natural sensory information. *IEEE Eng. Med. Biol. Mag.*, **18**, 91–98.
19 Donaldson, N., Rieger, R., Schuettler, M., and Taylor, J. (2008) Noise and selectivity of velocity-selective multi-electrode nerve cuffs. *Med. Biol. Eng. Comput.*, **46**, 1005–1018.
20 Schuettler, M., Seetohul, V., Taylor, J., and Donaldson, N. (2006) Velocity-selective recording from frog nerve using a multi-contact cuff electrode. *Conf. Proc. IEEE Eng. Med. Biol. Soc.*, **1**, 2962–2965.
21 Goodall, E.V., Lefurge, T.M., and Horch, K.W. (1991) Information contained in sensory nerve recordings made with intrafascicular electrodes. *IEEE Trans. Biomed. Eng.*, **38**, 846–850.
22 Lawrence, S.M., Dhillon, G.S., Jensen, W., Yoshida, K., and Horch, K.W. (2004) Acute peripheral nerve recording characteristics of polymer-based longitudinal intrafascicular electrodes. *IEEE Trans. Neural Syst. Rehabil. Eng.*, **12**, 345–348.
23 Lefurge, T., Goodall, E., Horch, K., Stensaas, L., and Schoenberg, A. (1991) Chronically implanted intrafascicular recording electrodes. *Ann. Biomed. Eng.*, **19**, 197–207.
24 Yoshida, K. and Stein, R.B. (1999) Characterization of signals and noise rejection with bipolar longitudinal intrafascicular electrodes. *IEEE Trans. Biomed. Eng.*, **46**, 226–234.
25 Ordelman, S.C.M.A., Kornet, L., Cornelussen, L., Buschman, R., H.P.J., and Veltink, P.H. (2010) An indirect component in the evoked compound action potential of the vagal nerve, *J. Neural Eng*, **7**, 9pp.
26 Tosato, M., Yoshida, K., Toft, E., Nekrasas, V., and Struijk, J.J. (2006) Closed-loop control of the heart rate by electrical stimulation of the vagus nerve. *Med. Biol. Eng. Comput.*, **44**, 161–169.
27 Tosato, M., Yoshida, K., Toft, E., and Struijk, J.J. (2007) Quasi-trapezoidal pulses to selectively block the activation of intrinsic laryngeal muscles during vagal nerve stimulation. *J. Neural Eng.*, **4**, 205–212.
28 Chen, S.A., Chiang, C.E., Tai, C.T., Wen, Z.C., Lee, S.H., Chiou, C.W., Ding, Y.A., and Chang, M.S. (1998) Intracardiac stimulation of human parasympathetic nerve fibers induces negative dromotropic effects: implication with the lesions of radiofrequency catheter ablation. *J. Cardiovasc. Electrophysiol.*, **9**, 245–252.
29 Quan, K.J., Van Hare, G.F., Biblo, L.A., Mackall, J.A., and Carlson, M.D. (2001) Endocardial stimulation of efferent parasympathetic nerves to the atrioventricular node in humans: optimal stimulation sites and the effects of digoxin. *J. Interv. Card Electrophysiol.*, **5**, 145–152.
30 Schauerte, P., Mischke, K., Plisiene, J., Waldmann, M., Zarse, M., Stellbrink, C., Schimpf, T., Knackstedt, C., Sinha, A., and Hanrath, P. (2001) Catheter stimulation of cardiac parasympathetic nerves in humans: a novel approach to the cardiac autonomic nervous system. *Circulation*, **104**, 2430–2435.
31 Bianchi, S., Rossi, P., Della Scala, A., and Kornet, L. (2009) Endocardial transcatheter stimulation of the AV nodal fat pad: stabilization of rapid ventricular rate response during atrial fibrillation in left ventricular failure. *J. Cardiovasc. Electrophysiol.*, **20**, 103–105.
32 Rossi, P., Bianchi, S., Barretta, A., Della Scala, A., Kornet, L., De Paulis, R., Bellisario, A., D'Addio, V., Pavaci, H., and

Miraldi, F. (2009) Post-operative atrial fibrillation management by selective epicardial vagal fat pad stimulation. *J. Interv. Card. Electrophysiol.*, **24**, 37–45.

33 Schwartz, P.J., De Ferrari, G.M., Sanzo, A., Landolina, M., Rordorf, R., Raineri, C., Campana, C., Revera, M., Ajmone-Marsan, N., Tavazzi, L., and Odero, A. (2008) Long term vagal stimulation in patients with advanced heart failure: first experience in man. *Eur. J. Heart. Fail.*, **10**, 884–891.

34 George, R., Salinsky, M., Kuzniecky, R., Rosenfeld, W., Bergen, D., Tarver, W.B., and Wernicke, J.F. (1994) Vagus nerve stimulation for treatment of partial seizures: 3. Long-term follow-up on first 67 patients exiting a controlled study. First International Vagus Nerve Stimulation Study Group. *Epilepsia*, **35**, 637–643.

35 Penry, J.K. and Dean, J.C. (1990) Prevention of intractable partial seizures by intermittent vagal stimulation in humans: preliminary results. *Epilepsia*, **31** (Suppl 2), S40–S43.

36 Ramsay, R.E., Uthman, B.M., Augustinsson, L.E., Upton, A.R., Naritoku, D., Willis, J., Treig, T., Barolat, G., and Wernicke, J.F. (1994) Vagus nerve stimulation for treatment of partial seizures: 2. Safety, side effects, and tolerability. First International Vagus Nerve Stimulation Study Group. *Epilepsia*, **35**, 627–636.

37 Rutecki, P. (1990) Anatomical, physiological, and theoretical basis for the antiepileptic effect of vagus nerve stimulation. *Epilepsia*, **31** (Suppl 2), S1–S6.

38 Ali, II., Pirzada, N.A., Kanjwal, Y., Wannamaker, B., Medhkour, A., Koltz, M.T., and Vaughn, B.V. (2004) Complete heart block with ventricular asystole during left vagus nerve stimulation for epilepsy. *Epilepsy Behav.*, **5**, 768–71.

39 Zhang, Y., Popovic, Z.B., Bibevski, S., Fakhry, I., Sica, D.A., Van Wagoner, D.R., and Mazgalev, T.N. (2009) Chronic vagus nerve stimulation improves autonomic control and attenuates systemic inflammation and heart failure progression in a canine high-rate pacing model. *Circ. Heart. Fail.*, **2**, 692–699.

40 McNeal, D.R. (1976) Analysis of a model for excitation of myelinated nerve. *IEEE Trans. Biomed. Eng.*, **23**, 329–337.

41 Fang, Z.P. and Mortimer, J.T. (1991) Selective activation of small motor axons by quasi-trapezoidal current pulses. *IEEE Trans. Biomed. Eng.*, **38**, 168–74.

42 Vuckovic, A., Tosato, M., and Struijk, J.J. (2008) A comparative study of three techniques for diameter selective fiber activation in the vagal nerve: anodal block, depolarizing prepulses and slowly rising pulses. *J. Neural Eng.*, **5**, 275–86.

43 Meier, J.H., Rutten, W.L., and Boom, H.B. (1995) Force recruitment during electrical nerve stimulation with multipolar intrafascicular electrodes. *Med. Biol. Eng. Comput.*, **33**, 409–17.

44 Meier, J.H., Rutten, W.L., Zoutman, A.E., Boom, H.B., and Bergveld, P. (1992) Simulation of multipolar fiber selective neural stimulation using intrafascicular electrodes. *IEEE Trans. Biomed. Eng.*, **39**, 122–34.

45 Veltink, P.H., van Alste, J.A., and Boom, H.B. (1988) Simulation of intrafascicular and extraneural nerve stimulation. *IEEE Trans. Biomed. Eng.*, **35**, 69–75.

46 Veltink, P.H., van Alste, J.A., and Boom, H.B. (1989) Multielectrode intrafascicular and extraneural stimulation. *Med. Biol. Eng. Comput.*, **27**, 19–24.

47 Yoshida, K. and Horch, K. (1993) Reduced fatigue in electrically stimulated muscle using dual channel intrafascicular electrodes with interleaved stimulation. *Ann. Biomed. Eng.*, **21**, 709–14.

48 Yoshida, K. and Horch, K. (1993) Selective stimulation of peripheral nerve fibers using dual intrafascicular electrodes. *IEEE Trans. Biomed. Eng.*, **40**, 492–4.

49 Janes, R.D., Brandys, J.C., Hopkins, D.A., Johnstone, D.E., Murphy, D.A., and Armour, J.A. (1986) Anatomy of human extrinsic cardiac nerves and ganglia. *Am. J. Cardiol.*, **57**, 299–309.

50 Mischke, K., Zarse, M., Schmid, M., Gemein, C., Hatam, N., Spillner, J., Dohmen, G., Rana, O., Saygili, E., Knackstedt, C., Weis, J., Pauza, D., Bianchi, S., and Schauerte, P. (2010) Chronic augmentation of the

parasympathetic tone to the atrioventricular node: a nonthoracotomy neurostimulation technique for ventricular rate control during atrial fibrillation. *J. Cardiovasc. Electrophysiol.*, **21**, 193–199.

51 Schauerte, P., Scherlag, B.J., Scherlag, M.A., Jackman, W.M., and Lazzara, R. (2000) Transvenous parasympathetic nerve stimulation in the inferior vena cava and atrioventricular conduction. *J. Cardiovasc. Electrophysiol.*, **11**, 64–69.

52 Gemein, C., Schauerte, P., Hatam, N., Rana, O.R., Saygili, E., Meyer, C., Eickholt, C., Schmid, M., Knackstedt, C., Zarse, M., and Mischke, K. (2009) Targeting of cardiac autonomic plexus for modulation of intracardiac neural tone. *Europace*, **11**, 1090–1096.

53 Zhang, Y. and Mazgalev, T.N. (2004) Achieving regular slow rhythm during atrial fibrillation without atrioventricular nodal ablation: selective vagal stimulation plus ventricular pacing. *Heart Rhythm*, **1**, 469–475.

54 Zhang, Y., Yamada, H., Bibevski, S., Zhuang, S., Mowrey, K.A., Wallick, D.W., Oh, S., and Mazgalev, T.N. (2005) Chronic atrioventricular nodal vagal stimulation: first evidence for long-term ventricular rate control in canine atrial fibrillation model. *Circulation*, **112**, 2904–2911.

55 Chen, P.S. and Tan, A.Y. (2007) Autonomic nerve activity and atrial fibrillation. *Heart Rhythm*, **4**, S61–S64.

56 Chiou, C.W., Chen, S.A., Kung, M.H., Chang, M.S., and Prystowsky, E.N. (2003) Effects of continuous enhanced vagal tone on dual atrioventricular node and accessory pathways. *Circulation*, **107**, 2583–2588.

57 Ogawa, M., Zhou, S., Tan, A.Y., Song, J., Gholmieh, G., Fishbein, M.C., Luo, H., Siegel, R.J., Karagueuzian, H.S., Chen, L.S., Lin, S.F., and Chen, P.S. (2007.) Left stellate ganglion and vagal nerve activity and cardiac arrhythmias in ambulatory dogs with pacing-induced congestive heart failure. *J. Am. Coll. Cardiol.*, **50**, 335–343.

58 Zhou, S., Jung, B.C., Tan, A.Y., Trang, V.Q., Gholmieh, G., Han, S.W., Lin, S.F., Fishbein, M.C., Chen, P.S., and Chen, L.S. (2008) Spontaneous stellate ganglion nerve activity and ventricular arrhythmia in a canine model of sudden death. *Heart Rhythm*, **5**, 131–139.

59 Swissa, M., Zhou, S., Tan, A.Y., Fishbein, M.C., Chen, P.S., and Chen, L.S. (2008) Atrial sympathetic and parasympathetic nerve sprouting and hyperinnervation induced by subthreshold electrical stimulation of the left stellate ganglion in normal dogs. *Cardiovasc. Pathol.*, **17**, 303–308.

60 Lewis, M.E., Al-Khalidi, A.H., Bonser, R.S., Clutton-Brock, T., Morton, D., Paterson, D., Townend, J.N., and Coote, J.H. (2001) Vagus nerve stimulation decreases left ventricular contractility *in vivo* in the human and pig heart. *J. Physiol.*, **534**, 547–552.

61 Sato, T., Kawada, T., Shishido, T., Sugimachi, M., Alexander, J. Jr., and Sunagawa, K. (1999) Novel therapeutic strategy against central baroreflex failure: a bionic baroreflex system. *Circulation*, **100**, 299–304.

62 Sato, T., Kawada, T., Sugimachi, M., and Sunagawa, K. (2002) Bionic technology revitalizes native baroreflex function in rats with baroreflex failure. *Circulation*, **106**, 730–734.

63 Sugimachi, M. and Sunagawa, K. (2005) Bionic cardiovascular medicine: functional replacement of native cardiovascular regulation and the correction of its abnormality. *IEEE Eng. Med. Biol. Mag.*, **24**, 24–31.

11
Bioelectronic Sensing of Insulin Demand

Matthieu Raoux, Guilherme Bontorin, Yannick Bornat, Jochen Lang, and Sylvie Renaud

This chapter presents potential design of a closed loop system in the context of therapy for diabetes, using integrated hardware as smart biohybrid sensors of glucose, hormones, and drugs. The devices process cellular action potentials, could be used for drug/toxicity screening, and act as integrated sensors for assessing insulin demand in the treatment of diabetes, mainly of type 1. In a closed loop system, the sensor output would be used as a control signal for driving an insulin pump. We present here novel results for the sensor and the conditioning/detection hardware tested on *in vitro* preparations of pancreatic endocrine cells, and propose an architecture for the closed loop system, inspired by previous systems we have designed for neuroscience applications.

11.1
Sensor Technologies and Cell Therapy in Diabetes: a Life-Long Debilitating Disease

Diabetes mellitus (DM) is a serious, debilitating, incurable, and costly disease that involves an electrically excitable microorgan, the pancreatic islets. This pathology poses the challenge for the application of the principles of bioelectronic hybrid systems in online investigation and the development of an artificial organ. In medical terms, DM is a disorder of carbohydrate and lipid metabolism characterized by impaired ability of the body to produce and/or respond to the hormone insulin produced by β-cells in pancreatic islets and thereby maintenance of proper levels of sugar (glucose) in the blood. The disease appears in two main forms: type 1 and type 2. The hallmarks of type 2 diabetes mellitus (T2DM) are deficient insulin secretion from pancreatic islets and insulin resistance in target tissues. In type 1 diabetes (T1DM), destruction of pancreatic β-cells usually leads to absolute insulin deficiency and concerns some 5–10% of all cases of DM. Diabetes causes a number of secondary complications such as heart disease, renal failure, and peripheral limb neuropathy that can lead to amputation, and in developed countries diabetes constitutes the leading cause of vision loss. A progression in the incidence of diabetes has been observed worldwide, including T1DM in the young, with a total of 220 million cases in 2010. In addition to the human burden caused by diabetes,

Biohybrid Systems: Nerves, Interfaces, and Machines, First Edition. Edited by Ranu Jung.

not only to the patients but also to their careers, the economic costs are enormous. Clearly, adapting newer technologies shall provide crucial human and economic benefits in treatment of diabetes.

Though lifestyle intervention (e.g., diet and exercise) combined with oral drug treatment remains the cornerstone of therapy, this will not suffice for many patients. Chronic insulin treatment is mandatory for patients afflicted by T1DM and the quality of long-term blood glucose control is decisive to reduce the occurrence of complications. The necessity of multiple daily hormone injections, mainly in T1DM, represents a major nuisance aggravated by repetitive blood sugar controls to ensure adequate therapy and avoid life-threatening hypoglycemia. Insulin therapy also poses the problem of difficult dosage management to adapt to diverse situations in everyday life and requires substantial and expensive paramedic assistance. These constraints led to the development of insulin pumps and the concept of the artificial pancreas in the 1980s. The systems in clinical use consist of an implantable insulin pump, continuous measurement internal glucose sensors that utilize enzyme-enhanced electrochemical reactions and software control algorithms. Despite the use of smart algorithms, the system still works essentially as an open loop mainly due to problems inherent to the sensor technology. Drawbacks include lag time in sensing and induced hypoglycemia, inability to cope with changes induced by exercise, puberty, and the menstrual cycle, just to mention a few. One reason for these drawbacks is the use of measurement of a single parameter, that is, glucose for the control algorithms, and the lack of reliability of electrochemical glucose sensors especially for detection of hypoglycemic levels.

The concept of an artificial pancreas has to be weighted in the light of cell therapy. Islet cell transplantation is indicated only in very restricted number of patients with unstable diabetes and its success is transient and burdened by immunosuppressive therapy. Although considerable progress has been achieved in producing islet cells from stem cells for a potential cure, general protocols are still not available, their applicability to man remains to be fully explored, and major hurdles remain, such as differentiation into specific lineages and the risk of inherent tumorigenicity leading potentially to teratomas and cancer. Even in the presence of successful stem cell therapy, the development of a bioartificial organ still has its place in transient use in critically ill patients, either as a diagnostic device for metabolic monitoring or as a screening module. It may, in fact, even benefit from a constant and well-defined source of cells.

11.2
The Biological Sensor: Function of β-Cells and Islet

Energy homeostasis of the human body relies on the precise sensing of nutrient levels and integration of hormonal signals to adapt to different physiological demands such as rest or exercise. The best-studied sensor is given by the islets of Langerhans, shaped for this purpose throughout evolution. Some 1–3 million islets are dispersed throughout the pancreas in humans accounting for about 2% of the total organ. They exhibit a size ranging from 50 to 500 μm and contain generally

1000–3000 cells, mainly insulin-secreting β-cells and also glucagon-releasing α- and few somatostatin-containing δ-cells. Islets precisely analyze blood nutrient levels such as glucose, due to the specific arrangement of their metabolic pathways, and provide exactly the required amount of insulin [1]. Pancreatic β-cells not only sense glucose levels but also take into account other nutrients, such as certain amino and fatty acids (Figure 11.1). The metabolism of these nutrients generates intracellular metabolic coupling factors such as ATP and the ensuing change in the ratio of ATP/ADP leads to the closure of ATP-sensitive K^+ channels (K_{ATP}), thereby depolarizing the cells from the resting potential (−70 mV) to above −55 mV. At this stage, the activation of voltage-dependent Ca^{2+} channels (Ca_v, mainly L-type) initiates action potentials further that are enhanced by voltage-dependent sodium channels (Na_v) [2] (Figure 11.1). Ca^{2+} influx through different ion channels subsequently triggers the release of insulin through exocytosis. Repolarization of the membrane potential is exerted by Ca^{2+}-activated potassium channels (BK) and to some degree by voltage-dependent K^+ channels (K_v). These ionic events are further tuned by the influence of intracellular second messengers, namely, cAMP generated by adenylyl cyclases that are regulated by hormones (such as incretins or adrenalin) as well as glucose itself [3]. The generation of action potentials (spikes) thus reflects the metabolic and signaling state. In addition, islets contain not only β- but also α-cells that react in response to glucose and incretins in a manner opposite to β-cells, that is, by reduced electrical activity [2]. This property may be used both for ratio measurements and for safer detection of hypoglycemia. Note that δ-cells respond to glucose mainly like β-cells [4].

Insulin secretion is also influenced by hormones, a modulation absolutely required for maintenance of homeostasis. The incretins GIP (glucose-dependent insulinotropic polypeptide) and GLP-1 (glucagon-like peptide-1) are secreted from intestinal cells upon the arrival of nutrients and sensitize the β-cell to the action of glucose [5]. They favor depolarization, whereas other hormones, such as

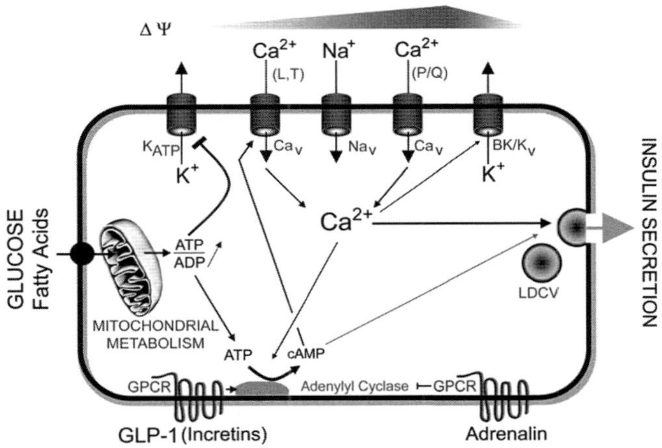

Figure 11.1 Glucose and hormones regulate insulin secretion via changes in the membrane potential (ΔΨ). For details, see text.

somatostatin or adrenaline, hyperpolarise β-cells to reduce insulin secretion during stress [6] (Figure 11.1). As all these hormones regulate secretion by impinging on the regulation of membrane conductances, electrical activity represents a coincidence readout of the state of β-cells or of α-cells and insulin demand. It therefore provides a highly informative signal. Thus, β-cells or islets provide the best sensor available, shaped continuously during evolution, both for continuous glucose monitoring and for pharmacological and toxicological screening.

11.3
Automated Islet Screening and Bioelectronic Sensor of Insulin Demand

We chose to use the integrative sensing properties of the islets by measuring and processing their electrical activity on silicon-based active electrode arrays. We condition and process analog voltages measured on each electrode, in order to detect and quantify the spiking activity, that itself testifies for the metabolic state of the islets. Using analog hardware, we can thus integrate "smart sensors" or "pixels" combining sensing and processing on a single substrate and guarantee online and real-time features for each measurement channel. Analog voltages from each channel of the array are conditioned in analog and online, using low-power/low-cost integrated circuits (ICs) able to quantify and decode the detected parameters in real time.

A deliverable device for automated functional islet screening is of interest in two major settings: first, drug development, and toxicity tests; second, quality control and toxicity testing in the frame of islet transplantations and generation of islets from stem cells. While the latter represents a small though highly active group in relation to the number of transplantations (approximately 400/year worldwide), diabetes drug discovery concerning islet function can be found in the portfolios of leading pharmaceutical companies (e.g., Eli Lilly, Novartis, Novo, Merck, and Servier) as well as small research-intensive companies (e.g., Incyte, VIA, Veroscience, Lexicon, and Arete) in line with the importance of diabetes. The highly informative readout of an automated long-term functional screening device occupies an important niche between in-depth single-cell work and large-scale single-time-point screening. Note that the latter often requires genetically engineered cells or loading of probes, thus considerably changing the environment or precluding long-term measurements.

Providing an autonomous, continuous hormone replacement therapy most closely adapted to physiology and responsive to different everyday life situations will provide a major advantage not only in terms of patients' life quality and expectancy [7, 8] but also in terms of cost effectiveness [9]. As outlined below, sensor technology is a major bottleneck in achieving this goal. The long-term goal, an implantable biomicroelectronic hybrid sensor of insulin demand, will provide a major advance in the attempt to approach normal lifestyles and reduce complications especially in patients suffering from T1DM. The major hurdle in this context is to prevent immunological attacks of implanted cells. Recent progress in this very active research field in material sciences has permitted production of semipermeable membranes that restrict access of immune components [10–12].

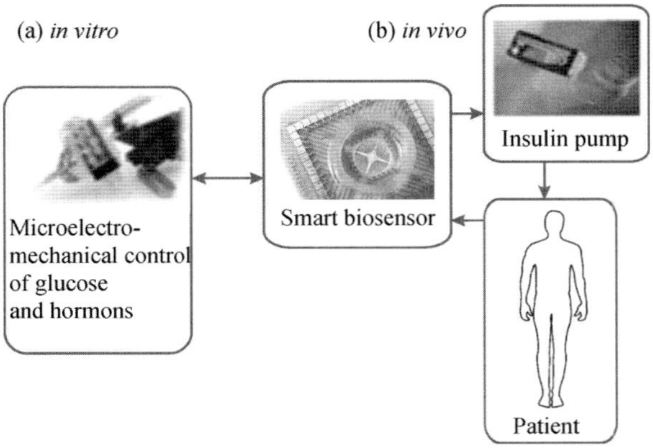

Figure 11.2 Closed loop embodiment of the smart biosensor, highlighting *in vitro* versus *in vivo* approaches. (a) *In vitro* setup: perfusions of glucose and hormones mimic the dynamics of insulin. (b) *In vivo* setup: glucose and levels of insulin are regulated by the patient via insulin injection from a pump.

11.4 Closed Loop Exploration *In Vitro*

This sensor and the conditioning/detection devices also serve to provide the proof of concept for closed loop functions. Figure 11.2 shows a potential *in vitro* closed loop configuration, as opposed to the *in vivo* configuration described above. As the device will not be implanted at the current stage, biological feedback control by glucose and levels of relevant hormones will not be feasible compared to a model implanted in animals or man. Therefore, buffers with dynamic glucose and hormone profiles will be delivered to the MEA device as "extracorporeal solutions" mimicking "absence" versus "presence" of appropriate concentrations of insulin (note in the reduced system here, insulin cannot diminish glucose as this action of the hormone is mediated mainly via liver, muscle, and fat cells). To calibrate drugs delivery profiles, preliminary simulations will be conducted to determine the dynamics required to stabilize the system.

11.5 Methods

11.5.1 Cultures and MEA

The classical approach for electrophysiological investigations of pancreatic islet cells is to utilize intracellular recording via patch-clamp [2]. However, this powerful method for the study of ion channels requires high expertise and allows recording of only one cell at a time. In addition, patch-clamp electrodes are quite invasive for cells, hence limiting the duration of recordings to several minutes. By using microelectrode arrays (MEAs) for the first time in this field, we developed a novel approach in which 60 microelectrodes are located outside the cells for extracellular, multicellular, noninvasive, and long-term recordings. Electrodes, were made from titanium nitride or platinum and had a diameter of 10–30 μm in order to limit the number of cells recorded by each channel. The minimal distance between two neighboring electrodes was 100 μm. This prevents recording of the same signal on several electrodes. The surface of the MEAs was coated with an extracellular matrix in order to improve cell adherence.

The ability to record from both clonal and primary cells was tested. Clonal INS-1E cells derived from rat insulinoma are the most commonly used cell line model of pancreatic β-cells. Islet cells were obtained from NMRI mice by enzymatic *in situ* digestion and subsequent isolation. The formation of islet-like clusters was observed after 2–3 days of culture on MEAs (Figure 11.3a).

Electrophysiological data were acquired with a MEA1060-Inv-BC-Standard amplifier (Multichannel Systems, Reutlingen, GE), sampled at 10 kHz, filtered at 0.01–3 kHz and analyzed with the MC Rack (Multichannel Systems) and Neuroexplorer (Nex Technologies) software.

11.5.2
Signal Conditioning and Spike Detection

As an initial step toward development of screening devices or closed loop insulin control systems, it is mandatory to design smart integrated circuits for processing online electrode signals. The planned use as a screening tool (multiple parallel channels), as well as the need for an implanted device, implies design constraints such as high signal-to-noise ratio, low power, low cost, and high integration density.

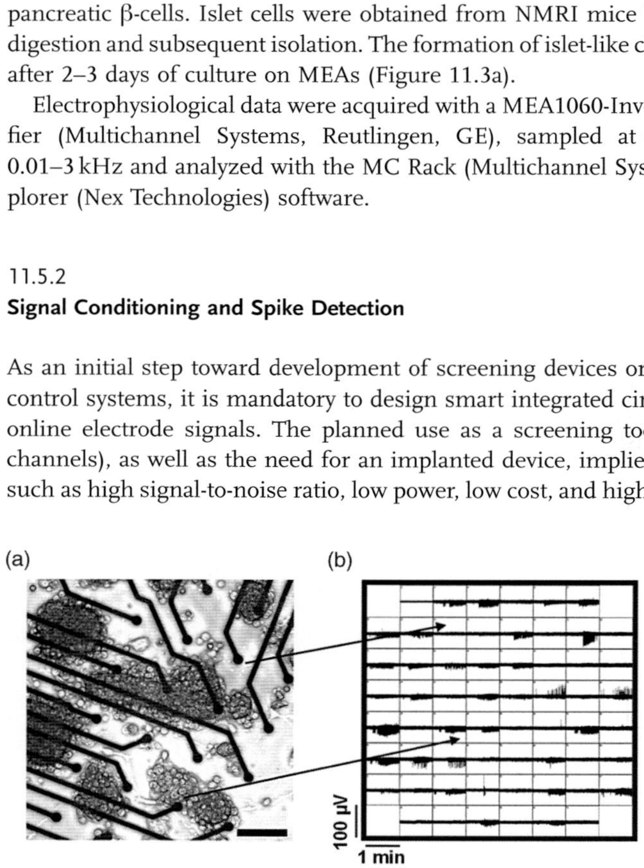

Figure 11.3 Culture and recording of islet cells on MEAs. (a) Mouse islet cells cultured for 2 days on the microelectrodes of an MEA (scale bar, 100 μm); (b) 60 recording channels corresponding to the electrodes of the MEA in (a). The upper arrow points to a biologically silent electrode (i.e., no cell on the electrode), whereas the lower one shows electrical signals generated by the cells growing on an electrode. We can observe that the signal amplitude is in tens of microvolts.

To have full control on device performances, ASICs were designed in full custom mode on standard (i.e., low cost) CMOS technology.

For each acquisition channel (microelectrode), a "pixel" was designed including a preamplifier and an activity detector. We use the term "pixel" by analogy with the APS (Analog Pixel Sensor) approach: APS image sensing devices are arrays of pixels where the photodetector is integrated on silicon together with its processing amplifier, in order to solve the speed and scalability issues of passive pixel sensors. In our case, the pixel approach not only helped solve the scale issue but also improved the signal-to-noise ratio.

Each pixel in our device consists of a preamplifier and an activity detector (Figure 11.4), and should be situated as close as possible to the electrode, under the MEA substrate, to maximize shielding from noise of the biological signal, which is limited to $\pm 50\,\mu V$ as explained in the previous section. The gain of the amplifier is set between 200 and 300, its cutoff frequency is between 10 and 30 kHz. An activity detector processes online the amplified signal. The detection method utilizes dynamic thresholding; the threshold is continuously adapted to the signal standard deviation. Its output is an asynchronous binary signal encoding the presence and duration of spikes. The detector details are presented in Section 11.6. Both the amplified signal and the detection signal are available as outputs of the pixel.

Note that we have not yet considered the microfluidic aspects that are important for biomedical applications. Other setups [13, 14] have successfully addressed this point, and our next series of prototypes, developed with collaborators [15], shall include microfluidic channels for drug delivery.

Our preliminary studies showed that it was possible to record and detect electrical activity on pancreatic β-cells. Results using the prototype setup providing real-time activity extraction using fixed and adaptive threshold methods are presented in the next section. In parallel, we are studying the hardware implementation of alternative detection methods based on wavelet transforms [16]. Preliminary results on noise reduction and detection mistakes are very encouraging. The setup can be adapted for real-time acquisition to suit the specific needs of closed loop applications.

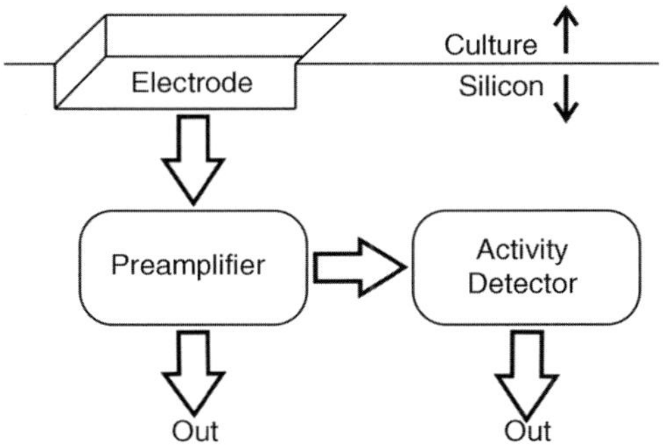

Figure 11.4 Integrated pixel for activity detection in cultures of pancreatic islet cells (see details in text).

11.6 Results

11.6.1 Recordings

Although MEAs have been widely used in neuroscience applications, we succeeded for the first time in the culture and the electrophysiological recording on MEAs of both clonal and primary endocrine β-cells (Figure 11.3) that are known to have much smaller electrical signals than neurons. The upper part of Figure 11.5a is a representative electrophysiological trace recorded on an active electrode of an MEA plated with mouse β-cells. At 37 °C, a rise in glucose concentration in the bath from 3 to 15 mM evoked typical cellular electrical events, namely, action potentials or spikes, in both primary (Figure 11.5a, upper trace) and INS-1E (not shown) β-cells. Classically, action potentials induced by glucose in β-cells are recorded using the perforated patch-clamp technique as shown in the lower part of Figure 11.5a. This technique consists of the technically demanding and rather time-consuming approach of sealing the tip of a glass microelectrode to the cell membrane and waiting for small perforations of the membrane patch by a pore-forming antibiotic previously introduced into the electrode. This configuration provides electrical access to the intracellular compartment and prevents washout of enzymes and coupling factors required for the response to glucose (see Section 11.2). Figure 11.5a (lower panel) also nicely demonstrates that β-cells depolarize only to around 0 mV, in stark contrast to the approximately 40 mV observed in neurons, thus explaining the small amplitude of extracellular recordings (Figure 11.5a, upper panel). Note that delays in the

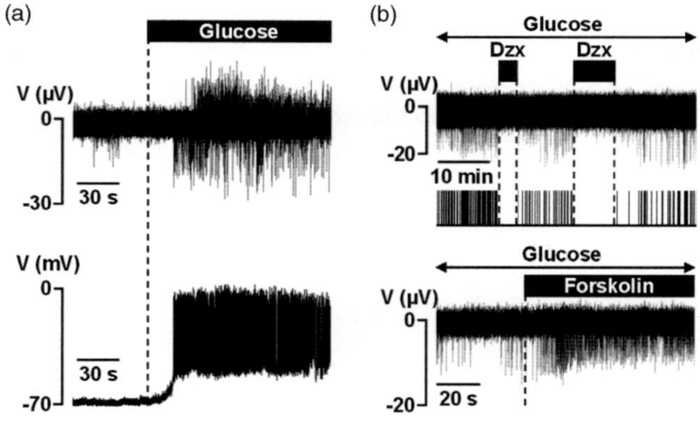

Figure 11.5 Representative electrophysiological recordings of primary and clonal β-cells. (a) Comparison between MEA (upper trace) and perforated patch-clamp (lower trace) for the recording of electrical signals of mouse β-cells stimulated with 15 mM of glucose. (b) Modulation of MEA-recorded electrical activities of INS-1E β-cells by 200 μM of the K_{ATP} channel opener diazoxide (Dzx, upper trace) and by 1 μM of the cAMP-raising agent forskolin (lower trace). Diazoxide and forskolin were applied in the presence of 15 mM of glucose. A raster plot presenting each spike by a vertical bar is given below the upper trace.

onset of glucose-evoked spike trains and spike frequencies were in the same order in both patch-clamp and MEA recordings (Figure 11.5a).

We confirmed that glucose-induced electrical activities recorded on MEAs involve the closure of K_{ATP} channels, as in the presence of high glucose (15 mM), action potentials were abolished in a reversible and reproducible manner by 200 μM of the K_{ATP} channel opener diazoxide (Figure 11.5b, upper trace). Taken together these elements indicate that MEAs are well suited for the study of islet cells in a noninvasive manner offering a much simpler approach compared to perforated patch-clamp. This device could also be used for pharmacological and long-term toxicological screening of drugs on islets cells. In addition, changes observed in the electrical pattern after glucose elevation constitute the basis for the development of a glucosensor.

The development of glucosensors, which also detect endocrine hormones regulating insulin secretion, remains a challenge. Thus, we decided to test whether the bioelectronic model was able to sense changes in the electrical activity of β-cells after stimulation with forskolin, a cAMP-raising agent that mimics the effects of the incretin hormones GLP-1 and GIP. In the presence of 15 mM of glucose, 1 μM of forskolin clearly increased the firing rate of INS-1E β-cells (Figure 11.5b, lower trace), demonstrating that a hybrid bioelectronic sensor based on this technology will take in account not only glucose but also other parameters crucial for the determination of insulin demand.

11.6.2 Adaptive Detection

Spiking and bursting patterns indeed depend on the glucose and hormone concentrations, and encode the demand in insulin or drug effects. Therefore, spike detection on extracellular recordings from β-cells is primordial to access such information. This detection may be complicated by superposition of multiple action potential from neighboring cells. Spikes can also be embedded in noise: the electrolyte–electrode interface generates a low-frequency potential difference between the solid electrode and the electrolyte solution. This potential varies spatially, from electrode to electrode, and temporally. For example, in recording from a gold-coated electrode in buffered saline solution, this DC offset can be as high as ±50 mV [17]. To process low-frequency noise fluctuations, we proposed an adaptive detection method in which the detection circuit (Figure 11.4) implements dynamic voltage thresholding. The circuit estimates in real time, and over a predefined time frame, the standard deviation (SD) of the signal. The threshold is set at a multiple N of SD, and thus dynamically updated (Figure 11.6a). Configuration parameters such as the length of the time window, the integer N, and the direction of the detected transition (up/down) are controlled by an external capacitance and by digital codes, respectively. Except for the control inputs, the detection circuit is entirely analog and was integrated using a low-cost MOS technology.

Figure 11.6b shows experimental results obtained by the integrated pixel and detectors processing in real time the signal recorded on a MEA from an islets culture. The signal was preamplified by the pixel IC as described in Section 11.5. The noise is

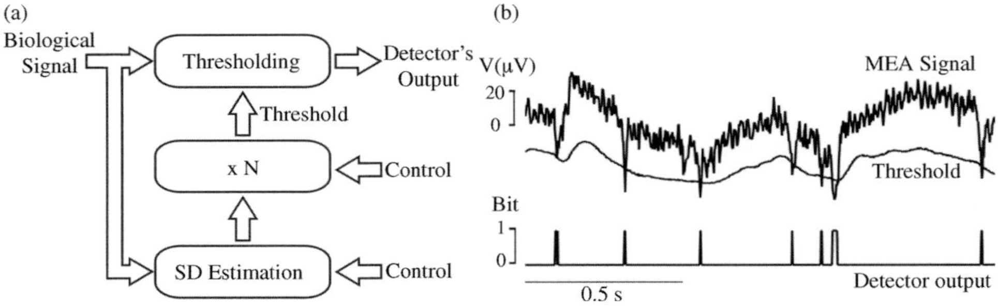

Figure 11.6 Spike detection in islets electrical activity: (a) architecture of the analog detection IC. The detector output and the control inputs are digital signals. (b) Activity detection of signals from primary β-cells recorded by MEAs. Voltages are scaled down to the value before amplification by the pixel IC. Upper plot: recorded signal. Middle plot: threshold voltage, as computed by the *SD estimation* and the *xN* subcircuits. Bottom plot: detector asynchronous output.

important and presents a fluctuating baseline. The adaptive threshold voltage contours the low-frequency noise (baseline). Detection is effective on down transitions of the signal, the time window is 100 ms, and N is set to 5. A high value for N allows avoiding detection faults that would otherwise appear, for example, at time 0.5 s, where the "low" spike occurrence was an artifact from electrical activity of a neighboring cell (not shown in the figure). For each detected spike, the width of the output pulse represents the duration of the spike, whose variation can provide useful information for postprocessing.

11.7 Conclusions

The proposed system is a major step in the development of an implantable sensor for insulin demand. As a screening device, it measures the major integrative function of islet cells, that is, electrical activity, without the requirement of prior probe loading or genetic manipulation and their inherent problems, but offering the possibility of long-term culture. It will be able to detect immediately acting compounds, and due to the possibility of long-term recordings, it can also detect slowly acting substances (up to days) in paired controls. Automatic signal interpretation qualifies it for medium- to high-throughput screening. In the long run, this can be extended to an implantable screening device recording effects of compounds in the whole animal.

Its conception as a future implantable sensor (Figure 11.7) for insulin demand uses the best sensor imaginable, that is, islets that have been shaped for this role during evolution and where all relevant signals tightly regulate action potentials. The project deals with the notoriously difficult issue of using living cells with limited lifetime and interfacing them to microelectromechanical systems. It exploits recent advances in hybrid bioelectronic interfaces using embedded ICs for signal

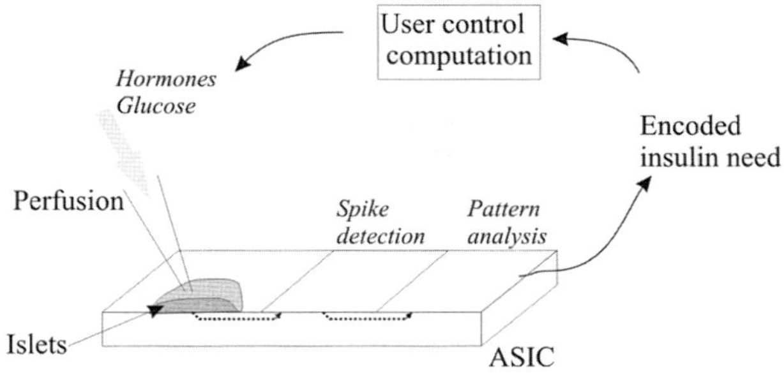

Figure 11.7 Implantable bioelectronic sensor for insulin need: the sensing, detection, and encoding functions are integrated into a single die using a low-power mixed analog/digital technology.

processing. As illustrated by several examples in this book, several advances in biohybrid systems are being made in the context of neurosciences, but we show here that the field can expand: dedicated cells such as islet cells have an unparalleled capacity for recognition and integration of relevant signals, thus providing a very attractive model for making biohybrid circuits to sense biochemical changes in the body. Because of the capability for miniaturization inherent to electronic systems, setups such as those described in this chapter offer major advantages compared to the design of chemical or photonic approaches for hybrid biosensors. Not only could implantable *in vivo* systems be developed specifically for insulin sensing but also may genetic modifications of the cells allow *in vitro* diagnostic systems for a wide variety of highly diverse compounds and signals.

References

1 Henquin, J.C., Dufrane, D., and Nenquin, M. (2006) Nutrient control of insulin secretion in isolated normal human islets. *Diabetes*, **55**, 3470–3477.

2 Braun, M., Ramracheya, R., Bengtsson, M., Zhang, Q., Karanauskaite, J., Partridge, C., Johnson, P.R., and Rorsman, P. (2008) Voltage-gated ion channels in human pancreatic beta-cells: electrophysiological characterization and role in insulin secretion. *Diabetes*, **57**, 1618–1628.

3 Tengholm, A. and Gylfe, E. (2009) Oscillatory control of insulin secretion. *Mol. Cell Endocrinol.*, **297**, 58–72.

4 Braun, M., Ramracheya, R., Amisten, S., Bengtsson, M., Moritoh, Y., Zhang, Q., Johnson, P.R., and Rorsman, P. (2009) Somatostatin release, electrical activity, membrane currents and exocytosis in human pancreatic delta cells. *Diabetologia*, **52**, 1566–1578.

5 Kim, W. and Egan, J.M. (2008) The role of incretins in glucose homeostasis and diabetes treatment. *Pharmacol. Rev.*, **60**, 470–512.

6 Iwanir, S. and Reuveny, E. (2008) Adrenaline-induced hyperpolarization of mouse pancreatic islet cells is mediated by G protein-gated inwardly rectifying potassium (GIRK) channels. *Pflugers Arch.*, **456**, 1097–1108.

7 Tamborlane, W.V., Beck, R.W., Bode, B.W., Buckingham, B., Chase, H.P., Clemons, R., Fiallo-Scharer, R., Fox, L.A., Gilliam, L.K., Hirsch, I.B., Huang, E.S., Kollman, C., Kowalski, A.J., Laffel, L., Lawrence, J.M., Lee, J., Mauras, N., O'Grady, M., Ruedy, K.J., Tansey, M., Tsalikian, E., Weinzimer, S.,

Wilson, D.M., Wolpert, H., Wysocki, T., and Xing, D. (2008) Continuous glucose monitoring and intensive treatment of type 1 diabetes. *N. Engl. J. Med.*, **359** 1464–1476.

8 Wentholt, I.M., Maran, A., Masurel, N., Heine, R.J., Hoekstra, J.B., and DeVries, J.H. (2007) Nocturnal hypoglycaemia in type 1 diabetic patients, assessed with continuous glucose monitoring: frequency, duration and associations. *Diabet. Med.*, **24**, 527–532.

9 Huang, E.S., O'Grady, M., Basu, A., Winn, A., John, P., Lee, J., Meltzer, D., Kollman, C., Laffel, L., Tamborlane, W., Weinzimer, S., and Wysocki, T. (2010) The cost-effectiveness of continuous glucose monitoring in type 1 diabetes. *Diabetes Care*, **33**, 1269–1274.

10 Kang, C., Vanoye, C.G., Welch, R.C., Van Horn, W.D., and Sanders, C.R. (2010) Functional delivery of a membrane protein into oocyte membranes using bicelles. *Biochemistry*, **49**, 653–655.

11 Gimi, B., Kwon, J., Liu, L., Su, Y., Nemani, K., Trivedi, K., Cui, Y., Vachha, B., Mason, R., Hu, W., and Lee, J.B. (2009) Cell encapsulation and oxygenation in nanoporous microcontainers. *Biomed. Microdevices*, **11**, 1205–1212.

12 Teramura, Y. and Iwata, H. (2010) Bioartificial pancreas microencapsulation and conformal coating of islet of Langerhans. *Adv. Drug. Deliv. Rev.*, **62**, 827–840.

13 Mohammed, J.S., Wang, Y., Harvat, T.A., Oberholzer, J., and Eddington, D.T. (2009) Microfluidic device for multimodal characterization of pancreatic islets. *Lab. Chip.*, **9**, 97–106.

14 Murua, A., Portero, A., Orive, G., Hernandez, R.M., Castro, M.D., and Pedraz, J.L. (2008) Cell microencapsulation technology: towards clinical application. *J. Control. Release*, **132**, 76–83.

15 Sloten, J., Verdonck, P., Nyssen, M., Haueisen, J., Spieth, S., Schumacher, A., Seidl, K., Hiltmann, K., Haeberle, S., McNamara, R., Dalley, J.W., Edgley, S.A., Ruther, P., and Zengerle, R. (2009) Robust microprobe systems for simultaneous neural recording and drug delivery, in *4th European Conference of the International Federation for Medical and Biological Engineering*, vol. **22** (eds R. Magjarevic and J.H. Nagel), pp. 2426–2430 Springer.

16 Nenadic, Z. and Burdick, J.W. (2005) Spike detection using the continuous wavelet transform. *IEEE Trans. Biomed. Eng.*, **52**, 74–87.

17 Robinson, D.A. (1968) The electrical properties of metal microelectrodes. *Proc. IEEE*, **56**, 1065–1071.

Index

Biohybrid Systems: Nerves, Interfaces, and Machines, First Edition. Edited by Ranu Jung.

o